SpringerBriefs in Mathematics

SpringerBriefs in Mathematics showcases expositions in all areas of mathematics and applied mathematics. Manuscripts presenting new results or a single new result in a classical field, new field, or an emerging topic, applications, or bridges between new results and already published works, are encouraged. The series is intended for mathematicians and applied mathematicians.

BCAM SpringerBriefs

BCAM *SpringerBriefs* aims to publish contributions in the following disciplines: Applied Mathematics, Finance, Statistics and Computer Science. BCAM has appointed an Editorial Board, who evaluate and review proposals.

Typical topics include: a timely report of state-of-the-art analytical techniques, bridge between new research results published in journal articles and a contextual literature review, a snapshot of a hot or emerging topic, a presentation of core concepts that students must understand in order to make independent contributions.

Please submit your proposal to the Editorial Board or to Francesca Bonadei, Executive Editor Mathematics, Statistics, and Engineering: francesca.bonadei@springer.com.

basque center for applied **mathematics**

More information about this series at http://www.springer.com/series/10030

Gary Chartrand · Cooroo Egan ·
Ping Zhang

How to Label a Graph

Gary Chartrand
Department of Mathematics
Western Michigan University
Kalamazoo, MI, USA

Cooroo Egan
Melbourne, VIC, Australia

Ping Zhang
Department of Mathematics
Western Michigan University
Kalamazoo, MI, USA

ISSN 2191-8198   ISSN 2191-8201   (electronic)
SpringerBriefs in Mathematics
ISBN 978-3-030-16862-9   ISBN 978-3-030-16863-6   (eBook)
https://doi.org/10.1007/978-3-030-16863-6

Mathematics Subject Classification (2010): 05C05, 05C10, 05C15, 05C45, 05C70, 05C78, 05C90

This Springer imprint is published by the registered company Springer Nature Switzerland AG
The registered company address is: Gewerbestrasse 11, 6330 Cham, Switzerland

Preface

Graph labelings have existed for well over a century. In fact, a graph G is often defined as a finite nonempty set V of objects called vertices and a set E of 2-element subsets of V called edges. This is further emphasized by expressing G as $G(V, E)$. If G has order n and size m, then the vertex set is often described as $V = V(G) = \{v_1, v_2, \ldots, v_n\}$ and the edge set as $E = E(G) = \{e_1, e_2, \ldots, e_m\}$. So, in a sense, the vertices of G are labeled as $v_1, v_2, \ldots, v_n$ and the edges of G are labeled as $e_1, e_2, \ldots, e_m$. The general topic of graph labelings is discussed in Chapter 1.

Graph labelings as an area of research in graph theory didn't really begin until the 1960s. Since then there have been numerous types of graph labelings introduced and studied. These have been described in a dynamic survey written by Joseph Gallian [18]. Indeed, prior to the current book, there have been five books written on graph labelings:

1. W. D. Wallis, *Magic Graphs*. Birkhäuser, Boston (2001).
2. M. Bača and M. Miller, *Super Edge-Antimagic Graphs*. BrownWalk Press, Boca Raton (2008).
3. A. M. Marr and W. D. Wallis, *Magic Graphs*, Second edition. Birkhäuser/ Springer, New York (2013).
4. M. Haviar and M. Ivaška, Vertex Labelings of Simple Graphs. Research and Exposition on Mathematics, Volume 34, Heldermann Verlag (2015).
5. S. C. López and F. A. Muntaner-Batle,*Graceful, Harmonious and Magic Type Labelings: Relations and Techniques*. SpringerBriefs in Mathematics. Springer, Cham (2017).

Three of these books have therefore dealt with the topic of magic labelings and the related antimagic labelings. The concept of magic labelings was introduced by Jiří Sedláček in the *Theory of Graphs and Its Applications*—a Symposium held in Smolenice, Czechoslovakia in June 1963. The proceedings of this Symposium concluded with a list of 31 problems, #27 of which included the following definition by Sedláček [41]:

A connected graph G is magic if there exists a real-valued function on the edge set of G with the properties that (i) distinct edges have distinct labels and (ii) the sum of labels of the edges incident with each vertex of G is the same constant.

Three years later, the American mathematician Bonnie M. Stewart [47] wrote a paper on magic graphs. This paper was followed by other papers by Stewart [48] and Sedláček [42] on magic labelings. Seven years after the Symposium in Smolenice, Anton Kotzig (who also attended the Symposium) and Alexander Rosa [29] proposed another type of magic labeling, which is now commonly called an edge-magic total labeling. For a graph G of order n and size m, an edge-magic total labeling of G is a bijective function $f : V(G) \cup E(G) \rightarrow [n+m] = \{1, 2, \ldots, n+m\}$ such that $f(uv) + f(u) + f(v)$ is the same constant for every edge uv of G. If $f(v)$ plus the sum of the labels of the edges incident with v is the same constant for every vertex v of G, then f is a vertex-magic total labeling of G.

The topic of magic labelings has been well covered, not only in the books mentioned above, but in numerous papers. Consequently, we will not discuss this area in the present book. Of the many graph labeling concepts and problems that have been introduced over the years, we have found some of particular interest to us that don't appear to have received such wide recognition. This will then be the emphasis of this book.

In 1966, Alexander Rosa [40] introduced a different view of vertex labeling, which later became known as a graceful labeling, when the vertices of a graph were labeled with distinct integers from a prescribed set that resulted in an edge labeling possessing a particular required property. This labeling is the primary topic in Chapter 2.

In 1980, Ronald Graham and Neil Sloane [20] introduced another vertex labeling called a harmonious labeling where the vertex labels are selected from the ring $\mathbb{Z}_m$ of integers modulo m, where m is the size of a graph, with the goal of producing an edge labeling in which edge labels are distinct. This is the primary topic in Chapter 3.

Because the labels chosen for vertices in a graph labeling have often been positive integers, it is not surprising that some conditions placed on certain vertex labels have a number-theoretic flavor—in particular, requirements on the greatest common divisor of the labels of specified pairs of vertices in the graphs. Consequently, the prime number decompositions of labels in these labelings play an important role. Furthermore, some of the labelings can be expressed in terms of subsets of certain sets of integers. This is the topic in Chapter 4.

As has already been noted, there have been several vertex labelings and edge labelings of graphs where each label is a positive integer and certain labels are added to produce a sum possessing a particular property. There are occasions when the resulting sums represent a coloring with a certain property. In these cases, the goal is often to minimize the largest label being used. This is the topic in Chapter 5.

Chapter 6 concerns vertex labelings of planar graphs with labels taken from the ring $\mathbb{Z}_3$ of integers modulo 3. In this case, the labels of the vertices on the boundary of each zone (or region) are added to produce a label for the zone with the goal of arriving at a particular sum in $\mathbb{Z}_3$. It is seen that there is a connection with these labelings and the famous Four Color Theorem.

Kalamazoo, MI, USA   Gary Chartrand
Melbourne, Australia   Cooroo Egan
Kalamazoo, MI, USA   Ping Zhang
February 2019

Contents

List of Figures

Chapter 1
Introduction

Graph labelings have been traced back to the 19th century when the famous British mathematician Arthur Cayley proved that there are n^{n-2} distinct labeled trees of order n (Cayley's Tree Formula). During the past several decades, this topic has become a popular area of research in graph theory. In this chapter, an introduction to labeled graphs and graph labelings is presented. A discussion of different interpretations of labeling graphs is given here as well.

1.1 Labeled Graphs

Arthur Cayley was a famous 19th century British mathematician. Among his many accomplishments, he was the first to define "group" in a modern way, namely as a set with a binary operation that satisfies certain laws. In fact, Cayley's theorem in algebra states that every group is isomorphic to a group of permutations. While Cayley is undoubtedly best known for his work in modern algebra, he also worked in graph theory and proved a theorem in graph theory that also bears his name.

A *tree* is a connected graph without cycles. Figure 1.1 shows all five trees of order 4 or 5, that is, having four or five vertices. None of these trees are labeled, that is, no vertex is assigned a label. Consequently, every tree of order 4 or 5 is isomorphic to one of the trees in Figure 1.1.

Suppose that two trees T_1 and T_2 of order n are labeled with the same set of n labels, say $1, 2, \ldots, n$. Then T_1 and T_2 are considered the same if they have the same set of edges (necessarily $n - 1$ edges). The distinct labeled trees of order 4 are shown in Figure 1.2. Therefore, while there are 16 distinct labeled trees of order 4, there are only two non-isomorphic trees of order 4. We saw in Figure 1.1 that there are only three non-isomorphic trees of order 5. However, there are 125 distinct labeled trees of order 5. Cayley proved the following in [5].

© The Author(s), under exclusive license to Springer Nature Switzerland AG 2019
G. Chartrand et al., *How to Label a Graph*, SpringerBriefs
in Mathematics, https://doi.org/10.1007/978-3-030-16863-6_1

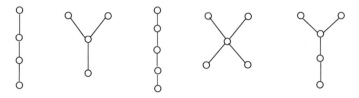

Fig. 1.1 The trees of order 4 or 5

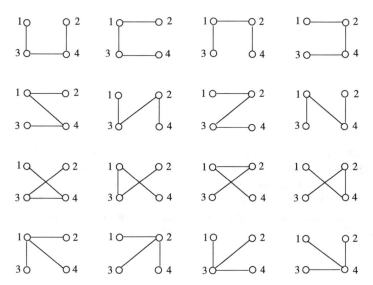

Fig. 1.2 The labeled trees of order 4

Theorem 1.1 (Cayley's Tree Formula) *For each positive integer n, there are n^{n-2} distinct labeled trees of order n.*

One of the best known proofs of this formula is due to Heinz Prüfer [36]. Indeed, there are many different proofs of Cayley's Tree Formula. John Moon [33] gave ten different proofs of this formula.

For a labeled graph G of order n, whose n vertices are labeled $1, 2, \ldots, n$, there are two possibilities for each of the $\binom{n}{2}$ pairs i, j of distinct vertices of G, namely either G contains the edge ij or it doesn't. This results in the following.

Observation 1.2 *For each positive integer n, there are $2^{\binom{n}{2}}$ distinct labeled graphs of order n.*

Consequently, there are $2^{\binom{3}{2}} = 2^3 = 8$ distinct labeled graphs of order 3. All of these graphs are shown in Figure 1.3. By Cayley's Tree Formula, three of these eight graphs are trees.

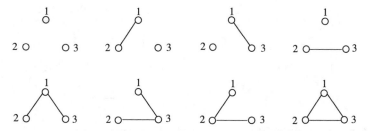

Fig. 1.3 The labeled graphs of order 3

1.2 Graph Labelings

Over the years, there have been different interpretations as to what is meant by labeling a graph. We employ the most general definition here. A *graph labeling* is an assignment of labels (elements of some set) to elements of a graph G, usually the vertices or the edges (or both) of G. If it is only the vertices of G that are labeled, then the resulting graph is a *vertex-labeled graph*. In the case of edges, the resulting graph is *edge-labeled*. Traditionally, a vertex labeling of a graph required distinct vertices to be assigned distinct labels. Typically, these labels were positive integers. So, if G has order n, the vertex labels were often chosen to be the elements of the set $[n] = \{1, 2, \ldots, n\}$. Similarly, an edge labeling of a graph G of size $m \geq 1$ assigns distinct labels from the set $[m] = \{1, 2, \ldots, m\}$ to the edges of G. More recently, however, there has been no requirement that the labels in a graph labeling be positive integers or be distinct. Figure 1.4 shows a vertex-labeled graph and an edge-labeled graph, where the labels in the first graph are nonnegative integers and the labels in the second graph are elements of $\mathbb{Z}_4$ (the integers modulo 4). In the first graph, two vertices are labeled the same; in the second graph, two pairs of edges are labeled the same.

In certain situations, it may be convenient for the labels to be colors, in which case, the labeling is a coloring. Traditionally, by a *coloring* of a graph G was meant an assignment of colors (elements of some set) to the vertices (or edges) of a graph G so that adjacent vertices (or edges) are assigned distinct colors. These are commonly called *proper colorings*. The major parameters dealing with proper colorings are the *chromatic number* $\chi(G)$ of a graph G, defined as the minimum number of colors in

Fig. 1.4 A vertex-labeled graph and an edge-labeled graph

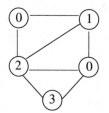

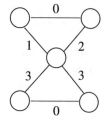

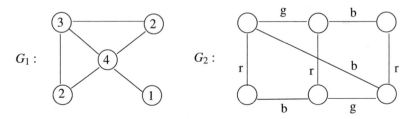

Fig. 1.5 A vertex-labeled graph G_1 and an edge-labeled graph G_2

a proper coloring of the vertices of G, and the *chromatic index* $\chi'(G)$, which is the minimum number of colors in a proper coloring of the edges of G. Over the years, this requirement has changed as well. For example, there has often been research dealing with *rainbow colorings* (both vertices and edges) where every two vertices (or edges) are required to be colored differently. That is, a rainbow coloring of the vertices of a graph G is equivalent to a traditional labeling of the vertices of G. Figure 1.5 shows a vertex-labeled graph G_1 where each vertex is labeled with its degree and an edge-labeled graph G_2 in which each edge of G_2 is assigned a color (r for red, b for blue, and g for green). Thus, G_2 is an edge-colored graph, where this coloring is being illustrated by means of an edge-labeled graph. Since no two adjacent edges of G_2 are colored the same, the coloring is a proper coloring.

It is also possible to have a graph in which all vertices (or edges) are colored the same. These are called *monochromatic colorings*. It is quite possible then to have a graph labeling in which vertices (or edges) are labeled the same and quite possible to have a coloring in which adjacent vertices (or adjacent edges) are colored the same. In any case, it must be made clear what conditions are being placed on any labeling or coloring being considered. In either case though, the reason for studying any labeling or coloring is to satisfy some prescribed conditions.

In the area of Ramsey Theory within graph theory, it is often the case that the edges of certain graphs (typically complete graphs) are arbitrarily colored with one of two colors (usually red or blue, resulting in a red-blue coloring) without any restriction whatsoever on the coloring, with the goal of looking for certain subgraphs all of whose edges are colored with a single color (that is, every edge of the subgraph is labeled the same), resulting in what is called a *monochromatic subgraph.*

More precisely, when we refer to a vertex labeling of a graph G, it is meant that there exists a function $f : V(G) \rightarrow S$ for some set S. The function f itself is called the *vertex labeling* of G and the elements $f(v)$ for $v \in V(G)$ are *vertex labels*. These labels may or may not be distinct and the set S may or may not be a set of positive integers. As we will see, the set S may be a set of nonnegative integers, S may be the set $\mathbb{Z}_k$ of integers modulo k for some integer $k \geq 2$, or S may be some other set. While it is customary for S to be a set whose elements can be added or subtracted, no such restriction on the elements of S is required in general.

Depending on the purpose for labeling the vertices of a graph, the labels themselves may be given other names. As mentioned above, if the labels are called colors, then the labeling is a coloring; while if the labels are called weights, then the labeling is a weighting. Furthermore, if the labels are values of some sort, then the resulting labeling is a valuation; while if the labels are numbers of some kind, then the labeling is a numbering. There are also occasions when labels may be referred to as costs. All of this applies as well to labeling edges.

In any graph theory problem involving a vertex labeling, there is always a purpose for labeling the vertices. Often the goal of a vertex labeling f of a graph G is its use in creating an edge labeling f' that has some prescribed property. Indeed, this process can be reversed by beginning with an edge labeling f of G and using it to generate a vertex labeling f' having a property of interest. Situations even more general than this are possible as we will see.

Over the years, a number of graph labelings have been introduced, some of which have led to curious and unexpected problems that have defied numerous attempts to solve them. In the chapters that follow, we will describe several of these labelings and resulting problems.

Chapter 2
Graceful Labelings

The vertex labeling that very well may have had the greatest influence on the development of graph labelings as a research area in graph theory is the graceful labeling. Some historical background of this concept is described in this chapter as well as its applications in communications networks. Much research on graceful labeling has been focused on determining which graphs have a graceful labeling and its connection with graph decompositions. Major results and conjectures on graceful labelings are presented, including the most famous conjecture on this topic dealing with trees.

2.1 Graceful Graphs

While graph theory is commonly considered to have begun in 1736, when Leonhard Euler solved and then generalized the famous Königsberg Bridge Problem, there was no real organization of this subject until 1936, when Dénes König wrote the first book [28] on the subject. Highlights of the first 200 years 1736–1936 of graph theory were described in a book by Norman Biggs, E. Keith Lloyd, and Robin Wilson [4] in 1976. The first two centuries of graph theory primarily consisted of some isolated results and contributions to recreational mathematics. Graph theory, however, never started developing into a theoretical and research area of mathematics until after World War II. This development essentially began during the 1950s and 1960s. The second book on the theory of graphs was written by the French mathematician Claude Berge [3] in 1958. The next such book was published in 1962 and written by Oystein Ore [34]. The book by Ore was the first written in English, as König's was written in German and Berge's in French.

Late in the 1950s and well into the 1960s, conferences were organized whose major emphasis was on graph theory. One of these conferences, called the Symposium on the Theory of Graphs and Its Applications, took place in Smolenice

© The Author(s), under exclusive license to Springer Nature Switzerland AG 2019
G. Chartrand et al., *How to Label a Graph*, SpringerBriefs
in Mathematics, https://doi.org/10.1007/978-3-030-16863-6_2

(in Czechoslovakia at the time) during June 17–20, 1963. The editors of the proceedings [14] of this conference wrote:

> It can easily be proved (a somewhat lengthy proof is given in this volume . . .) that the theory of graphs is growing steadily and has applications in so many fields of science as not many other branches have done before.

> A further important feature of graph theory is that a good part of its problems is understandable to a wide circle of readers. For this reason, for many years it has found its way into books on recreational mathematics with the object of attracting the interest of young students to mathematics.

> These facts led to the arrangement of Symposia on graph theory in Budapest 1958, Halle/Saale 1960 and, in 1963, in Smolenice.

The 1963 Symposium in Smolenice was quite likely the first major conference on graph theory. It featured 37 participants (24 from Czechoslovakia and 13 from outside the country). Among these participants were some of the best known mathematicians at that time known for their contributions to graph theory. This 4-day conference began with six half-day sessions, followed by discussions of problems on graph theory. The proceedings of this conference (see [14]) contained much of the mathematics presented in the lectures as well as 31 problems that were presented there. Problem #25, presented by Gerhard Ringel, stated a conjecture of his.

Ringel's Conjecture *For every tree T of size m, the complete graph K_{2m+1} can be decomposed into $2m + 1$ copies of T*

Later, Anton Kotzig presented an even stronger conjecture.

Kotzig's Conjecture *For every tree T of size m, the complete graph K_{2m+1} can be cyclically decomposed into $2m + 1$ copies of T*

We will return to these conjectures soon.

Other conferences on the theory of graphs followed soon afterward, including one in Rome during July 5–9, 1966 and another in the United States in Kalamazoo, Michigan in 1968. Graph theory conferences became commonplace in the 1970s and thereafter. One of the participants of the Rome conference was Alexander Rosa who presented a paper [40] that would not only have a major influence on the area of graph labelings but would be directly related to the previously mentioned Ringel–Kotzig conjecture. In his paper, Rosa introduced four vertex labelings which he referred to as α- *valuations*, β-*valuations*, σ-*valuations*, and ρ-*valuations*. It was the β-valuation that would play a significant role in the study of labelings. This labeling would acquire a new name in 1972 when a β-valuation was referred to as a *graceful labeling* by Solomon Golomb [19]. A possible application of these labelings suggested by Golomb can be described as follows:

> Suppose that a connected graph of order n and size m represents a communications network having n terminals and m interconnections between terminals. Distinct identifying numbers are to be assigned to each terminal in a way that uniquely identifies the interconnections by assigning to each interconnection the absolute value of the difference of the numbers assigned to its two end-terminals. If the goal is to minimize the largest number assigned to a terminal, then the resulting problem is directly related to graceful labelings.

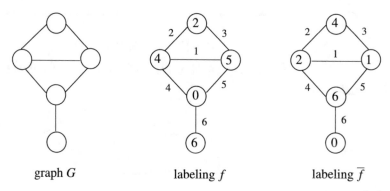

graph G labeling f labeling $\overline{f}$

Fig. 2.1 A graceful graph

Let G be a nonempty graph of order n and size m. The graph G has a *graceful labeling* f if the vertices of G can be assigned distinct elements from the set $[0, m] = \{0, 1, \ldots, m\}$ in such a way that the resulting edge labeling f', defined by $f'(uv) = |f(u) - f(v)|$ for every edge uv of G, results in distinct edges receiving distinct labels. That is, the function $f : V(G) \rightarrow [0, m]$ is injective and the function $f' : E(G) \rightarrow [m] = \{1, 2, \ldots, m\}$ is bijective. If a graph G admits a graceful labeling, then G itself is called a *graceful graph*. For example, the graph G of order 5 and size 6 shown in Figure 2.1 is graceful. A graceful labeling f of G is also shown in Figure 2.1.

Since $|f(u) - f(v)|$ is the distance $d(f(u), f(v))$ between the integers $f(u)$ and $f(v)$ on the real line, another way to determine whether a given graph is graceful is suggested by this observation. Let G be a nonempty graph of order n and size m and consider the points $0, 1, 2, \ldots, m - 1, m$ on the real line. The distance $d(i, j) = |i - j|$ between any two of these points $i, j \in [0, m]$, $i \neq j$, is one of the integers $1, 2, \ldots, m$. Suppose that there exists a subset V of $[0, m]$ with $|V| = n$ for which there is a set E of m pairs of distinct elements of V where the distances between the integers in these pairs are the elements of $[m]$. Let H be the graph with vertex set V and edge set E. If $H \cong G$, then G is graceful. If there is no such set V of n elements of $[0, m]$ with this property, then G is not graceful. For example, consider the graph G shown in Figure 2.2. This graph G has order $n = 6$ and size $m = 7$. Consider the points $0, 1, \ldots, 7$ on the real line (also shown in Figure 2.2). Now consider the subset $V = \{0, 1, 2, 4, 6, 7\}$ of the set $\{0, 1, \ldots, 7\}$ consisting of these six integers and the set

$$E = \{\{1, 2\}, \{2, 4\}, \{1, 4\}, \{2, 6\}, \{1, 6\}, \{0, 6\}, \{0, 7\}\}$$

of seven pairs of elements of V. The distances between the integers in these seven pairs are $1, 2, \ldots, 7$. The graph H with vertex set V and edge set E is isomorphic to G. Therefore, G is graceful.

If f is a graceful labeling of a graph G of order n and size m, then the vertex labeling $\overline{f}$ of G defined by

$$\overline{f}(v) = m - f(v) \text{ for each vertex } v \text{ of } G$$

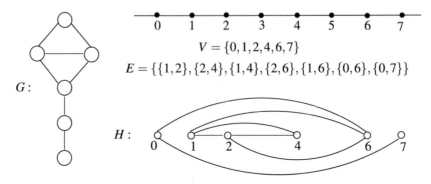

Fig. 2.2 Showing that a graph G is graceful

Fig. 2.3 The Petersen graph
is graceful

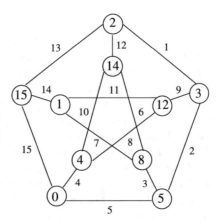

is called the *complementary labeling* of f. The edge labeling $\overline{f}'$ defined by

$$\overline{f}'(uv) = |\overline{f}(u) - \overline{f}(v)| = |(m - f(u)) - (m - f(v))| = |f(u) - f(v)| = f'(uv)$$

shows that $\overline{f}$ is also a graceful labeling of G. The complementary graceful labeling $\overline{f}$ of the graph G of Figure 2.1 is also shown in the figure. Another example of a graceful graph is the famous Petersen graph. A graceful labeling of this graph is shown in Figure 2.3.

There is a slightly different but equivalent way to determine whether a graph G is graceful. Again, suppose that G is a graph of order n and size m. Consider the complete graph K_{m+1} whose $m + 1$ vertices are labeled $0, 1, \ldots, m$. If two vertices of K_{m+1} are labeled i and j with $i > j$, then the edge joining these two vertices is labeled $i - j$. The graph G is graceful if G is isomorphic to a subgraph of K_{m+1} whose m edges are labeled 1 through m in K_{m+1}. This is illustrated in Figure 2.4 for the graph G of Figure 2.1.

In the vertex-labeled and edge-labeled complete graph K_{m+1} just described, there are $m + 1 - i$ edges labeled i for $1 \leq i \leq m$. Determining whether a given graph G of order n and size m is graceful by this method requires us to construct a subgraph

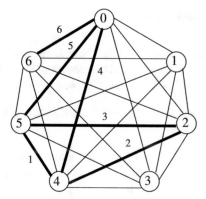

Fig. 2.4 Showing that the graph G of Figure 2.1 is graceful

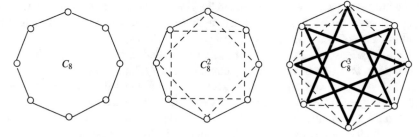

Fig. 2.5 The graphs C_8, C_8^2, and C_8^3

of K_{m+1} that is isomorphic to G by selecting the single edge labeled m, one of the two edges labeled $m-1$, one of the three edges labeled $m-2$, and so on. There are other edge-labeled graphs of interest to us with edges labeled 1 through m, but where there are an equal number of edges of each label.

Let G be a connected graph of order n. By the kth *power* G^k of G is meant that graph where $V(G^k) = V(G)$ such that $uv \in E(G^k)$ if $1 \leq d_G(u, v) \leq k$, that is, if the distance between u and v is at most k in G. The graph G^2 is called the *square* of G and G^3 is called the *cube* of G. Observe for $n \geq 7$ that the diameter of the n-cycle C_n is $\lfloor n/2 \rfloor \geq 3$ and that for each such integer n, the graph C_n^3 is 6-regular. More generally, for integers n and k with $n \geq 2k + 1 \geq 5$, $\text{diam}(C_n) = \lfloor n/2 \rfloor \geq k$ and C_n^k is $2k$-regular. In fact, if $n = 2k + 1$, then $C_n^k = C_{2k+1}^k = K_{2k+1}$, which, of course, is $2k$-regular. Furthermore, there are exactly n pairs of vertices in C_n whose distance in C_n is i where $1 \leq i \leq k$. Each such pair of vertices in C_n is joined by an edge in C_n^k. If we label each edge uv of C_n^k by $d_{C_n}(u, v)$, then C_n^k contains exactly n edges labeled i for every integer i with $1 \leq i \leq k$. This is illustrated in Figure 2.5 for $n = 8$ and $k = 2, 3$. The first graph is C_8, the second is its square C_8^2, and the third is its cube C_8^3 in which each edge uv is labeled with $d_{C_8}(u, v)$. In Figure 2.5, a dashed line joins two vertices u and v with $d_{C_8}(u, v) = 2$, while a bold line joins two vertices u and v with $d_{C_8}(u, v) = 3$.

2.2 Cyclic Decompositions

One feature of graceful graphs G is that there is always a complete graph of a particular order that can be decomposed into copies of G—not only "decomposed" but "cyclically decomposed" into copies of G, which we are about to describe.

Theorem 2.1 *If G is a graceful graph of size m, then the complete graph K_{2m+1} can be cyclically decomposed into G.*

Proof Since G is graceful, there is a graceful labeling of G such that the vertices of G are labeled from a subset of $\{0, 1, \ldots, m\}$ so that the induced edge labels are $1, 2, \ldots, m$. Let $V(K_{2m+1}) = \{v_0, v_1, \ldots, v_{2m}\}$ where the vertices of K_{2m+1} are arranged cyclically in a regular $(2m + 1)$-gon and labeled in clockwise order. Denote the resulting $(2m + 1)$-cycle by $C = (v_0, v_1, v_2, \ldots, v_{2m-1}, v_{2m}, v_0)$. For each integer i with $0 \leq i \leq m$, a vertex labeled i in the graceful labeling of G is placed at v_i in K_{2m+1}. Every edge of G is drawn as a straight-line segment in K_{2m+1}, denoting the resulting copy of G in K_{2m+1} by G_1. Hence, $V(G_1) \subseteq \{v_0, v_1, \ldots, v_m\}$.

Each edge $v_s v_t$ of K_{2m+1} ($0 \leq s, t \leq 2m$) is labeled $d_C(v_s, v_t)$, where then $1 \leq d_C(v_s, v_t) \leq m$. Consequently, K_{2m+1} contains exactly $2m + 1$ edges labeled i for each i ($1 \leq i \leq m$) and G_1 contains exactly one edge labeled i for each integer i with $1 \leq i \leq m$. Whenever an edge of G_1 is rotated clockwise through an angle of $2\pi k/(2m + 1)$ radians, where $1 \leq k \leq 2m$, an edge of the same label is obtained. The subgraph obtained by rotating G_1 through a clockwise angle of $2\pi k/(2m + 1)$ radians is denoted by G_{k+1}. Then $G_{k+1} \cong G$ and a cyclic decomposition of K_{2m+1} into $2m + 1$ copies of G results. $\square$

Theorem 2.1 is illustrated for the graceful graph K_3 in Figure 2.6, where a graceful labeling of K_3 is shown. Since K_3 has size $m = 3$ and $2m + 1 = 7$, there is a cyclic decomposition of K_7 into seven copies of K_3. The first copy G_1 of K_3 is also shown in Figure 2.6 with solid lines. By rotating G_1 clockwise through an angle of $2\pi/7$ radians, a second copy G_2 of K_3 is obtained. The copy G_2 is also shown in Figure 2.6, where the edges of G_2 are drawn with dashed lines. Rotating G_1 five more times clockwise through angles of $2\pi/7$ radians produces a cyclic decomposition of K_7 into seven copies of K_3.

Suppose that, instead of starting with the $(2m + 1)$-cycle C and the complete graph $K_{2m+1} = C_{2m+1}^m$, as described in the proof of Theorem 2.1, we were to begin with a cycle C_p for some integer $p \geq 2m + 1$ and consider the graph C_p^m, where $V(C_p^m) = \{v_0, v_1, v_2, \ldots, v_{p-1}\}$, whose vertices are arranged cyclically in a regular p-gon and labeled in clockwise order. If we proceed as in the proof but rotate G a total of $p - 1$ times, then we obtain the following.

Theorem 2.2 *Let G be a graceful graph of order n and size m and let p be an integer with $p \geq 2m + 1$. Then the graph C_p^m can be cyclically decomposed into p copies of G.*

Theorem 2.2 is illustrated for K_3 and $p = 9$ in Figure 2.7.

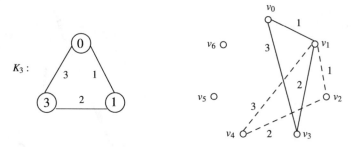

Fig. 2.6 Cyclically decomposing K_7 into seven copies of K_3

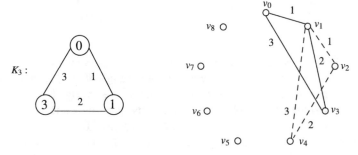

Fig. 2.7 Cyclically decomposing C_9^3 into nine copies of K_3

Of course, cyclically decomposing C_p^m, where $p \geq 2m + 1$, into p copies of a graph G of size m, according to Theorem 2.2, depends on G being graceful. Not all graphs are graceful, however. One such non-graceful graph is the 5-cycle C_5.

Proposition 2.1 *The 5-cycle C_5 is not graceful.*

Proof Assume, to the contrary, that C_5 is graceful. Since the size of C_5 is 5, the five vertices of C_5 must be labeled with five of the six labels $0, 1, \ldots, 5$ in such a way that the labels of the edges of C_5 are $1, 2, 3, 4, 5$. That is, exactly one of the six integers $0, 1, \ldots, 5$ is not a vertex label and exactly three edge labels are odd. Thus, either (1) three of the vertex labels are odd integers and two are even or (2) three of the vertex labels are even integers and two are odd. Since either the given graceful labeling or its complementary graceful labeling satisfies (1), we may assume that the given labeling satisfies (1). There are three odd labels and two even labels among the five edge labels. An edge uv has an odd label if and only if the labels of u and v are of opposite parity. Suppose that x and y are the two vertices of C_5 having even labels. If x and y are adjacent on C_5, then only two edges of C_5 have odd labels. If x and y are not adjacent, then four edges of C_5 have odd labels. In either case, this is impossible. $\qquad\qquad\qquad\Box$

Since C_5 is not graceful, it therefore does not follow from Theorem 2.1 that the complete graph K_{11} can be cyclically decomposed into copies of C_5. Nevertheless, it can be so decomposed. The graph $G_1 = C_5$ in Figure 2.8 has its edges

Fig. 2.8 A cyclic
decomposition of K_{11} into
copies of C_5

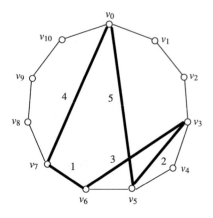

labeled $1, 2, 3, 4, 5$, where an edge $v_s v_t$ of G_1 ($0 \leq s, t \leq 10$) is labeled $i \in [5]$ and $d_{C_{11}}(v_s, v_t) = i$. By rotating G_1 clockwise through an angle of $2\pi k / 11$ radians, where $1 \leq k \leq 10$, another copy G_{k+1} of C_5 is obtained where an edge labeled i in G_1 is rotated into another edge labeled i in G_{k+1}. Consequently, this results in a cyclic decomposition $\{G_1, G_2, \ldots, G_{11}\}$ of K_{11} into 11 copies of C_5.

We have seen for every graceful graph G of size m that there is a cyclic decomposition of K_{2m+1} into $2m + 1$ copies of G. Yet, there may be a cyclic decomposition of K_{2m+1} into $2m + 1$ copies of a non-graceful graph G of size m. Indeed, we just saw that the non-graceful graph C_5 has this property. If a graph G can be drawn in the complete graph K_{2m+1} containing the Hamiltonian cycle $C = (v_0, v_1, v_2, \ldots, v_{2m-1}, v_{2m}, v_0)$ such that for each integer i with $1 \leq i \leq m$, there is an edge xy of G such that $d_C(x, y) = i$, then K_{2m+1} can be cyclically decomposed into $2m + 1$ copies of G. If G can be drawn in the "first half" of K_{2m+1}, namely, in the complete subgraph K_{m+1} with vertex set $\{v_0, v_1, v_2, \ldots, v_m\}$, then G is graceful; while if this is not possible, then G is not graceful. This observation was also made by Rosa [40] in his original paper on the subject, dealing with another valuation he introduced.

Since C_{11} has diameter 5, it follows that $C_{11}^5 = K_{11}$. Not only can C_{11}^5 be cyclically decomposed into C_5, this is also the case for C_{13}^5 and C_{15}^5, which is illustrated in Figure 2.9. This follows from three observations: (i) $5 - 2 + 3 + 1 + 4 = 11$, (ii) $2 + 3 + 4 - 1 + 5 = 13$, and (iii) $1 + 2 + 3 + 4 + 5 = 15$.

The so-called *bow tie graph* $B = 2K_2 \vee K_1$ (the join of $2K_2$ and K_1) shown in Figure 2.10 is another connected graph of order 5 that is not graceful. In fact, this graph B, the 5-cycle C_5, and the complete graph K_5 are the only connected non-graceful graphs of order 5.

Proposition 2.2 *The bow tie graph B of Figure 2.10 is not graceful.*

Proof Assume, to the contrary, that B is graceful. Since the size of B is 6, the five vertices of B must be labeled with five of the seven labels $0, 1, \ldots, 6$ in such a way that the labels of the edges of B are $1, 2, \ldots, 6$. That is, the edges of B must be labeled with three even integers and three odd integers. If the three vertices of a triangle in

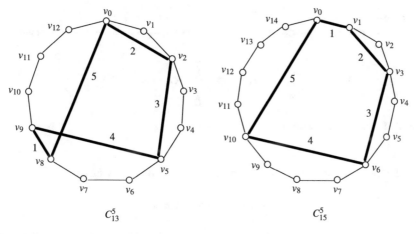

Fig. 2.9 Illustrating cyclic decompositions of C_{13}^5 and C_{15}^5 into copies of C_5

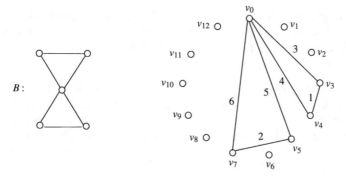

Fig. 2.10 A cyclical decomposition of K_{13} into copies of the bow tie graph

B are all labeled with even integers or are all labeled with odd integers, then no edge of this triangle is labeled with an odd integer. If this is not the case, then only one vertex of a triangle is labeled with an even integer or only one vertex is labeled with an odd integer. In either case, exactly two edges of the triangles are labeled with an odd integer. Therefore, regardless of how the six vertices of B are labeled, an even number of edges of B are labeled with odd integers. This is impossible. □

As with the graph C_5, even though B is not graceful, there is a cyclic decomposition of K_{13} into copies of B as shown in Figure 2.10.

As we have seen, if G is a graceful graph of size m, then every graph C_p^m where $p \geq 2m + 1$ can be cyclically decomposed into copies of G. However, even if G is a non-graceful graph of size m, then it may occur that C_p^m can be cyclically decomposed into copies of G for *some* integers $p \geq 2m + 1$. For a given connected graph G of size m, graceful or not, there are numerous questions here regarding those integers $p \geq 2m + 1$ for which the graph C_p^m is decomposable into copies of G.

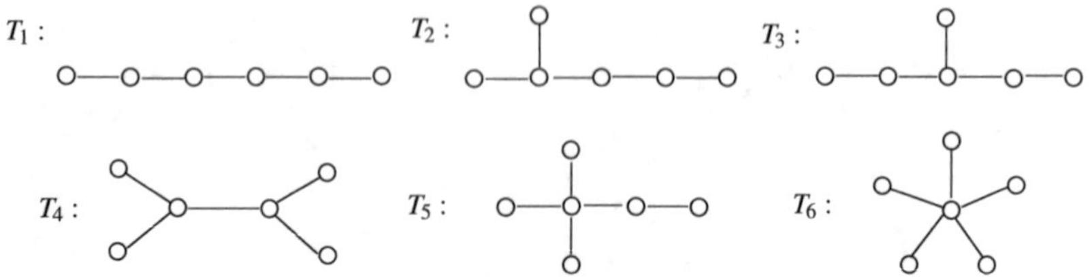

Fig. 2.11 The six trees of order 6

The major question remaining here, however, is the following:

Which graphs are graceful?

There are several well-known classes of graphs where it has been determined which members of the class are graceful.

1. The complete graph K_n $(n \geq 2)$ is graceful if and only if $n \leq 4$ [19].
2. The cycle C_n is graceful if and only if $n \equiv 0$ (mod 4) or $n \equiv 3$ (mod 4) [40].
3. Every complete bipartite graph $K_{r,s}$ is graceful [40].
4. Every wheel $W_n = C_{n-1} \vee K_1$ is graceful [16].

Wheels were shown to be graceful by Roberto Frucht [16], who is well known for solving the problem introduced in König's book [28] that for every finite group Γ, there exists a graph whose automorphism group is isomorphic to Γ. Paul Erdős and Robin Wilson [11] proved that almost all graphs are not graceful in 1977. That is, if $\mathcal{G}_n$ is the set of all graphs of order n and $\mathcal{R}_n$ is the set of all graceful graphs of order n, then $\lim_{n \to \infty} \frac{|R_n|}{|\mathcal{G}_n|} = 0$.

2.3 Graceful Trees

In order for a graph G of order n and size m with $n = m + 1$ to be graceful, each of the integers $0, 1, 2, \ldots, m$ must be used (exactly once) for a vertex label. If the graph G is connected, then G is a tree. Figure 2.11 shows all six trees of order 6.

Each tree of order 6 is graceful. One way this can be verified is by beginning with the complete graph K_6 whose vertices are labeled $0, 1, \ldots, 5$, where the edge joining the vertices labeled i and j $(i > j)$ is itself labeled, with the label $i - j$. See Figure 2.12. Each of the trees $T_1, T_2, \ldots, T_6$ of order 6 can be found in K_6 where the labels of the five edges in each tree are $1, 2, \ldots, 5$. See Figure 2.12. Therefore, all trees of order 6 are graceful.

A *double star* is a tree of diameter 3. Thus, if T is a double star, then T contains exactly two vertices that are not end-vertices, called the *central vertices* of T. For integers s and t with $s, t \geq 2$ and $s + t = m + 1$, let $S_{s,t}$ denote the double star of size m whose central vertices have degrees s and t, respectively. Not only are the path P_7,

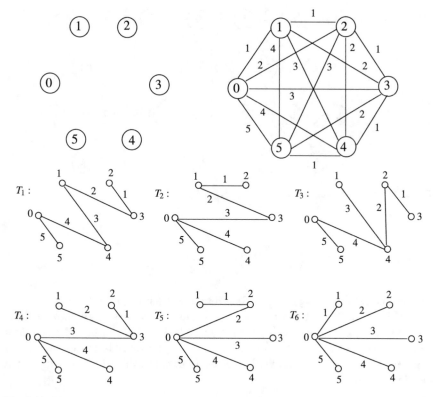

Fig. 2.12 The graph K_6 and the graceful trees $T_1, T_2, \ldots, T_6$

the star $K_{1,6}$, and the double star $S_{3,4}$ of size 6 graceful, every path, star, and double star is graceful.

Theorem 2.3 *Every path, star, and double star is graceful.*

Proof First, we show that the path $P_{m+1} = (v_0, v_1, v_2, \ldots, v_{m-1}, v_m)$ of size $m \geq 1$ is graceful. Consider the following vertex labeling f of P_{m+1}:

$$f(v_i) = \begin{cases} \frac{i}{2} & \text{if } i \text{ is even} \\ m - \frac{i-1}{2} & \text{if } i \text{ is odd.} \end{cases}$$

This is a graceful labeling of P_{m+1}. Consequently, every path is graceful.

Next, we show that the star $K_{1,m}$ of size $m \geq 1$ is graceful. Let v be the vertex of degree m in $K_{1,m}$ and let $v_1, v_2, \ldots, v_m$ be the remaining vertices. Consider the following vertex labeling f of $K_{1,m}$:

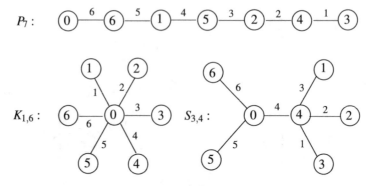

Fig. 2.13 Graceful labelings of P_7, $K_{1,6}$, and $S_{3,4}$

$$f(x) = \begin{cases} 0 & \text{if } x = v \\ i & \text{if } x = v_i \text{ for } 1 \leq i \leq m. \end{cases}$$

This is a graceful labeling of $K_{1,m}$. Therefore, every star is graceful.

Finally, for integers s and t with $s, t \geq 2$ and $s + t = m + 1$, we show that the double star $S_{s,t}$ of size $m \geq 3$ is graceful. Let u be the vertex of degree s and let v be the vertex of degree t in $S_{s,t}$. Let $u_1, u_2, \ldots, u_{s-1}$ be the end-vertices adjacent to u and $v_1, v_2, \ldots, v_{t-1}$ the end-vertices adjacent to v. Consider the following vertex labeling f of $S_{s,t}$:

$$f(x) = \begin{cases} 0 & \text{if } x = u \\ m + 1 - i & \text{if } x = u_i \text{ for } 1 \leq i \leq s - 1 \\ m + 1 - s & \text{if } x = v \\ i & \text{if } x = v_i \text{ for } 1 \leq i \leq t - 1. \end{cases}$$

Since this labeling is graceful, every double star is graceful.   □

Figure 2.13 illustrates the graceful labelings for the path P_7, star $K_{1,6}$, and double star $S_{3,4}$, as described in Theorem 2.3.

Indeed, it is believed that every nontrivial tree is graceful. This is the most famous conjecture in the area of graceful labelings.

The Graceful Tree Conjecture   *Every nontrivial tree is graceful.*

The Graceful Tree Conjecture is credited jointly to Anton Kotzig (Rosa's advisor) and Gerhard Ringel [38], who with J. W. T. Youngs is known for completing the solution of the famous Heawood Map Coloring Problem [39], namely, the problem of determining the largest chromatic number of a graph that can be embedded on an orientable surface of positive genus (see [9, p. 219]).

The Graceful Tree Conjecture has been verified for several classes of trees, including the following:

1. Every caterpillar is graceful [40].
2. Every tree with at most four end-vertices is graceful [24].
3. Every tree of order at most 35 is graceful [12, 18].
4. Every tree of diameter at most 5 is graceful [23].

Joseph Gallian's dynamic survey [18] of graph labelings contains a vast list of periodically updated results on the subject.

Chapter 3
Harmonious Labelings

In this chapter, we discuss vertex labelings where each label is selected from $\mathbb{Z}_m$ for a graph of size m. This labeling then gives rise to an edge labeling where the label of an edge is the sum of the labels of its incident vertices. The primary interest is when the edge labels are distinct.

3.1 Harmonious Graphs

The ring $\mathbb{Z}_3$ of integers modulo 3 consists of the three elements 0, 1, 2. When adding the $\binom{3}{2} = 3$ pairs of elements of $\mathbb{Z}_3$, we obtain all elements of $\mathbb{Z}_3$:

$$0 + 1 = 1, 0 + 2 = 2, 1 + 2 = 0.$$

The ring $\mathbb{Z}_6$ consists of the six elements 0, 1, 2, 3, 4, 5. It turns out that there exists a set S of four elements of $\mathbb{Z}_6$ such that when the $\binom{4}{2} = 6$ pairs of elements of S are added, the six resulting sums are precisely the elements of $\mathbb{Z}_6$. In particular, for $S = \{0, 1, 2, 4\}$, we have

$$0 + 1 = 1, 0 + 2 = 2, 0 + 4 = 4, 1 + 2 = 3, 1 + 4 = 5, 2 + 4 = 0.$$

This set S consisting of four elements of $\mathbb{Z}_6$ is referred to as an *additive basis* for $\mathbb{Z}_6$ in that every element of $\mathbb{Z}_6$ can be expressed as the sum of two distinct elements of S. Since $\binom{5}{2} = 10$, this brings up the question as to whether $\mathbb{Z}_{10}$ has an additive basis consisting of five elements of $\mathbb{Z}_{10}$. The answer to this question is *no*. Indeed, for every integer $m = \binom{n}{2}$, where $n \geq 5$, there is no additive basis for $\mathbb{Z}_m$ consisting of n elements of $\mathbb{Z}_m$.

The observations mentioned above suggest a vertex labeling concept. For example, consider the complete graph K_3 whose vertices are labeled with the elements 0, 1, 2 of $\mathbb{Z}_3$. From this vertex labeling, an edge labeling of K_3 is produced where the label of an edge is the sum (in $\mathbb{Z}_3$) of the labels of its two incident vertices and the three edge

© The Author(s), under exclusive license to Springer Nature Switzerland AG 2019
G. Chartrand et al., *How to Label a Graph*, SpringerBriefs
in Mathematics, https://doi.org/10.1007/978-3-030-16863-6_3

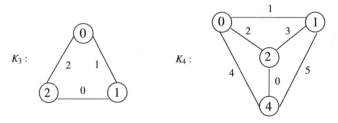

Fig. 3.1 Harmonious labelings of K_3 and K_4

labels are distinct. This is shown in Figure 3.1. In the same figure, K_4 is also shown where the four vertices are labeled with the distinct elements of $S = \{0, 1, 2, 4\}$ of $\mathbb{Z}_6$. As we already noted, there is no way to label the five vertices of K_5 in such a way that the resulting labels of the ten edges of K_5 are the ten elements $0, 1, \ldots, 9$ of $\mathbb{Z}_{10}$.

More generally, let G be a connected graph of order n and size m. Even though $m \geq n - 1$, let us assume for the present that $m \geq n$. The graph G is called a *harmonious graph* if there exists a set S of n distinct elements of $\mathbb{Z}_m$ with the property that the elements of S can be assigned to the vertices of G in such a way that each edge uv of G is labeled with the sum in $\mathbb{Z}_m$ of the labels of u and v and the resulting edge labels are distinct, namely, $0, 1, \ldots, m - 1$. Such a labeling of G is called a *harmonious labeling*. Hence, for a connected graph G of order n and size m with $m \geq n$, a vertex labeling f of G is a *harmonious labeling* if $f : V(G) \rightarrow \mathbb{Z}_m$ is injective and the resulting edge labeling $f' : E(G) \rightarrow \mathbb{Z}_m$ defined by $f(uv) = f(u) + f(v)$ is bijective. A connected graph is therefore *harmonious* if it admits a harmonious labeling. Thus, K_3 and K_4 are harmonious, while for each integer $n \geq 5$, the graph K_n is not.

Since a connected graph of order n and size m with $m = n - 1$ is a tree, the condition $m \geq n$ restricts ourselves to connected graphs that are not trees. The topic of harmonious trees will be discussed in the next section. Consequently, to determine whether a connected graph of order n and size m with $m \geq n$ is harmonious, it is necessary to determine whether there exists a harmonious labeling of G, that is, whether the vertices of G can be labeled with n distinct elements of $\mathbb{Z}_m$ such that the resulting edge labels of G are the elements of $\mathbb{Z}_m$. For example, for the famous Petersen graph, which is a cubic graph of order 10 and size 15, there are 10 distinct elements of $\mathbb{Z}_{15}$ that can be assigned to the vertices of the Petersen graph so that the resulting edge labels are the elements of $\mathbb{Z}_{15}$. A harmonious labeling of the Petersen graph is shown in Figure 3.2.

Additive bases and the resulting concept of harmonious labelings of graphs were introduced by Ronald Graham and Neil Sloane [20] in 1980, who, at that time, were members of Bell Laboratories. Graham is known for his many contributions to discrete mathematics and Sloane is probably best known for introducing the Online Encyclopedia of Integer Sequences.

Let $m \geq 3$ be an integer and consider a regular m-gon whose m vertices are labeled cyclically in clockwise order with the elements $0, 1, \ldots, m - 1$ of $\mathbb{Z}_m$. Every two

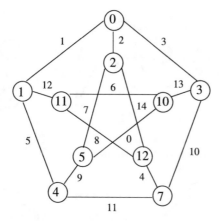

Fig. 3.2 The Petersen graph is harmonious

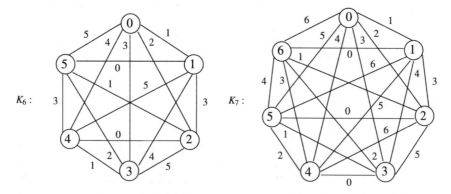

Fig. 3.3 The complete graph K_m labeled with elements of $\mathbb{Z}_m$ for $m = 6$ and $m = 7$

vertices of this m-gon are then joined by a straight-line segment. For two vertices of the m-gon labeled i and j, the edge joining them is labeled $i + j = k \in \mathbb{Z}_m$. All edges in this complete graph K_m that are parallel to an edge labeled k are also labeled k. See Figure 3.3 for the complete graphs K_6 and K_7.

Let G be a connected graph of order n and size m with $m \geq n$. If the vertices and edges of K_m are labeled with the elements of $\mathbb{Z}_m$ as indicated above and there exists one edge of each parallel class that results in a graph isomorphic to G, then G is harmonious. Figure 3.4 shows three different harmonious graphs of size 6, no two of which have the same order.

We have mentioned that only the complete graphs K_n with $n \leq 4$ are harmonious, which is exactly the same situation for graceful complete graphs. We also saw in Chapter 2 that "half" of the cycles are graceful. This too is true for harmonious cycles, although it is not the same half.

Theorem 3.1 *The cycle C_n, $n \geq 3$, is harmonious if and only if n is odd.*

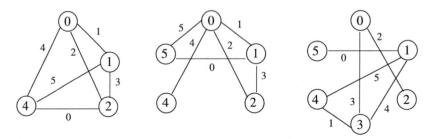

Fig. 3.4 Three harmonious graphs of size 6 having different orders

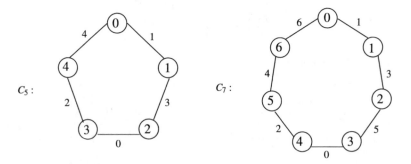

Fig. 3.5 Harmonious labelings of C_5 and C_7

Proof We have already seen that $C_3 = K_3$ is harmonious. For an odd integer $n \geq 5$, let $C_n = (v_0, v_1, \ldots, v_{n-1}, v_0)$, where the vertices of C_n are listed in clockwise order. We now label the vertex v_i, $0 \leq i \leq n - 1$, with the element $i \in \mathbb{Z}_n$ (see Figure 3.5). Beginning with the edge $v_0 v_1$ and proceeding clockwise about C_n, the edge labels are then $1, 3, 5, \ldots, n - 2, 0, 2, 4, \ldots, n - 1$. This is a harmonious labeling of C_n.

Next, let $n \geq 4$ be an even integer. Assume, to the contrary, that there is a harmonious labeling of C_n, where, say, v_i is labeled $a_i \in \mathbb{Z}_n$ for $0 \leq i \leq n - 1$. Hence, $\mathbb{Z}_n = \{a_0, a_1, \ldots, a_{n-1}\}$. Thus,

$$\sum_{i=0}^{n-1} a_i = \sum_{i=0}^{n-1} i = \binom{n}{2} \pmod{n}.$$

Since each vertex of C_n contributes its label as a term in the sum of two edge labels, it follows that

$$\binom{n}{2} = 2\sum_{i=0}^{n-1} a_i = 2\sum_{i=0}^{n-1} i = 2\binom{n}{2} \pmod{n}.$$

Consequently, $2\binom{n}{2} \equiv \binom{n}{2} \pmod{n}$ and so $\binom{n}{2} \equiv 0 \pmod{n}$. However, then, $n \mid \frac{n(n-1)}{2}$, which is impossible since $(n - 1)/2$ is not an integer. □

3.2 Harmonious Trees

As we mentioned earlier, we are interested in those connected graphs of order n and size m for which a harmonious labeling is possible. Thus far, we have only discussed connected graphs with $m \geq n$. However, it is possible that $m = n - 1$ for a connected graph G, in which case G is a tree. It is harmonious trees that we discuss now. Since the labels used in a harmonious labeling of a tree T of order n and size $m = n - 1$ are the elements of $\mathbb{Z}_m$, it is impossible to assign n distinct elements of $\mathbb{Z}_m$ to the n vertices of T. To rectify this situation in their definition of harmonious labeling, Graham and Sloane allowed exactly two vertices of a tree to be assigned the same label.

It is easy to see that all paths and stars are harmonious. Indeed, a harmonious labeling of a path of even order is much like that of a cycle. For paths $P_n = (v_0, v_1, \ldots, v_{n-1})$ of odd order $n = 2k + 1$ and size $n - 1 = 2k$, the vertices are labeled so that the induced edge labels are $f'(v_i v_{i+1}) = i + k$ for $0 \leq i \leq k - 1$ and $f'(v_i v_{i+1}) = i - k$ for $k \leq i \leq 2k - 1$. By defining the label $f(v_0) = 0$, the labels $f(v_i)$ of the remaining vertices v_i, $1 \leq i \leq n - 1$, are all uniquely determined. Furthermore, all vertex labels are distinct except that $f(v_1) = f(v_{n-1}) = k$. Harmonious labelings of P_n for $n = 6, 7, 8, 9$ and the star $K_{1,6}$ are shown in Figure 3.6.

Many other classes of trees have been shown to be harmonious. Among them are the following:

1. Every caterpillar is harmonious [20].
2. Every tree of order at most 31 is harmonious [12, 18].

In fact, Graham and Sloane [20] made the following conjecture, which parallels the famous conjecture on graceful labelings of trees.

The Harmonious Tree Conjecture *Every nontrivial tree is harmonious.*

If f is a harmonious labeling of a connected graph G of size m, then the labeling $\overline{f} : V(G) \rightarrow \mathbb{Z}_m$ of G defined by $\overline{f}(v) = m - 1 - f(v)$ for each $v \in V(G)$ is called

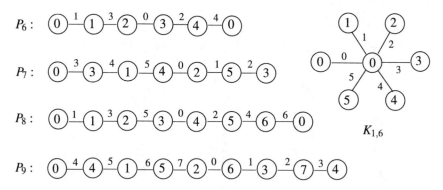

Fig. 3.6 Harmonious labelings of P_n for $n = 6, 7, 8, 9$ and $K_{1,6}$

the *complementary labeling* of f and is also a harmonious labeling of G. If T is a nontrivial tree of size m and the label $i \in \mathbb{Z}_m$ is repeated in a harmonious labeling f of T, then $m - 1 - i$ is the repeated label in $\overline{f}$. If m is odd and $i = (m - 1)/2$, then the repeated label is the same in both f and $\overline{f}$.

For a nontrivial tree T of size m, let $h(T)$ denote the number of elements of $\mathbb{Z}_m$ that can be repeated in some harmonious labeling of T. Therefore, $0 \le h(T) \le m$ for every tree T of size m. Of course, if the Harmonious Tree Conjecture is true, then $1 \le h(T) \le m$. In fact, if T is a tree of even size m and the Harmonious Tree Conjecture is true, then $h(T)$ is even and $2 \le h(T) \le m$. Figure 3.7 shows four trees, each of size m where m is 2, 3, or 5. For each of these four trees T, it follows that $h(T) = m$. That is, every element of $\mathbb{Z}_m$ is the repeated label in some harmonious labeling of T. This brings up several questions.

Problem 3.1 Does there exist a class of trees T for which $h(T) = 1$?

Problem 3.2 For which pairs k, m of positive integers with $k \le m$, does there exist a tree T of size m for which $h(T) = k$?

We mentioned earlier that to determine whether a connected graph G of order n and size m with $m \ge n$ is harmonious, we can begin with a regular m-gon whose vertices are labeled cyclically in clockwise order with the elements $0, 1, 2, \ldots, m - 1$ of $\mathbb{Z}_m$. Every two vertices of this m-gon, labeled by i and j say, are joined by a straight-line segment labeled $i + j \in \mathbb{Z}_m$. This produces the complete graph K_m, each of whose

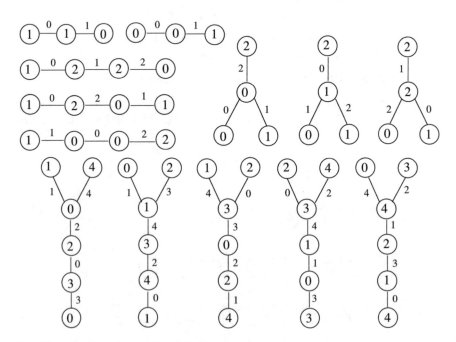

Fig. 3.7 Trees T for which $h(T) = |E(T)|$

edges is labeled with an element of $\mathbb{Z}_m$. Then two edges of K_m have the same label if and only if they are parallel. Thus, the edge set of K_m is partitioned into m parallel classes. To determine whether the graph G is harmonious, we then need to determine if one edge can be selected from each of the m parallel classes to arrive at a graph isomorphic to G.

There is only one harmonious graph of size 2, namely, the path P_3 (see Figure 3.8). There are two harmonious trees of size 3, namely, the path P_4 and the star $K_{1,3}$. There is only one other harmonious graph of size 3. If we begin with a regular 3-gon, then there is one edge in each parallel class, that is, K_3 is the remaining harmonious graph of size 3. There are three trees of size 4, all harmonious. To determine which connected graphs of size 4 that are not trees but are also harmonious, we begin with a regular 4-gon, whose vertices are labeled cyclically with the elements 0, 1, 2, 3 of $\mathbb{Z}_4$ in clockwise order (see Figure 3.8). There is only one edge in the parallel class joining the vertices labeled 0 and 2 and one edge in the parallel class joining the vertices labeled 1 and 3. So, these two edges must be selected. There are two edges in the parallel class joining the vertices labeled 0 and 1 (as well as in the parallel class joining the vertices labeled 1 and 2). Only one harmonious graph results here, shown in Figure 3.8.

In addition to the six trees of size 5, the harmonious graphs of size 5 consist of the four graphs shown in Figure 3.9.

Each harmonious graph of size $m \geq 6$ that is not a tree can be found in a similar manner, namely, by beginning with a regular m-gon whose vertices are labeled cyclically in clockwise order with the elements 0, 1, ..., $m-1$ of $\mathbb{Z}_m$ and selecting one edge from each of the m parallel classes. For $m = 6$, one harmonious graph obtained is $G = (K_4 - e) + K_2$, which is disconnected. A harmonious labeling $f : V(G) \to \mathbb{Z}_6$ of G can be defined by assigning the labels 1 and 4 to the two vertices of degree 3

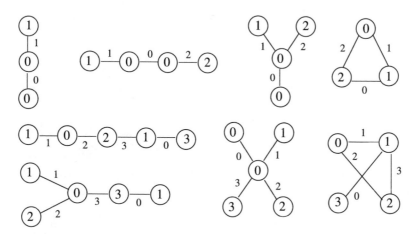

Fig. 3.8 Harmonious graphs of sizes 2, 3, and 4

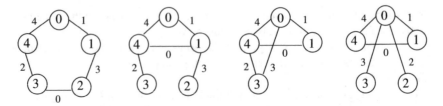

Fig. 3.9 Harmonious graphs of size 5 that are not trees

in G, the labels 0 and 2 to the two vertices of degree 2 in G, and the labels 3 and 5 to the two end-vertices in G. This harmonious graph has two components. This brings up the problem of determining a harmonious disconnected graph with a specified number of components.

Chapter 4
Prime Labelings

Unlike the situation with graceful labelings and harmonious labelings, where a vertex labeling induces an edge labeling, the emphasis in this chapter is on graphs whose vertices can be labeled with positive integers so that the labels of every two adjacent vertices have a particular number-theoretic property. We also see that some of these labelings can be looked at in a set-theoretic manner.

4.1 Prime Graphs

Each of the vertex labelings described in the two preceding chapters has led to an edge labeling possessing a property of interest. In this chapter, once again our interest lies with graphs possessing certain vertex labelings, but in many instances there is no corresponding edge labeling. The labelings we present in this chapter have either a number-theoretic or a set-theoretic flavor to them. In each instance, however, prime numbers play a role, either directly or indirectly.

Let G be a graph of order n. By a *prime labeling* of G is meant a vertex labeling of G with the distinct integers in the set $[n] = \{1, 2, \ldots, n\}$ such that the labels of every two adjacent vertices of G are relatively prime (also called *coprime*). That is, a prime labeling is a bijective function $f : V(G) \to [n]$ such that $\gcd(f(u), f(v)) = 1$ for each pair u, v of adjacent vertices of G. If there exists a prime labeling of G, then G is called a *prime graph*. This concept was originated by Roger Entringer (see [18]) around 1980. For example, the graph Q_3 of the cube is a prime graph. A prime labeling of this graph is shown in Figure 4.1. While adjacent vertices must have relatively prime labels, nonadjacent vertices may have relatively prime labels as well.

There are certain classes of graphs, the prime graphs of which can be determined quite quickly. One of these is the class of cycles.

Observation 4.1 *Every cycle is a prime graph.*

Proof Let $C_n = (v_1, v_2, \ldots, v_n, v_1)$, where $n \geq 3$. Define the labeling $f : V(C_n) \to [n]$ by $f(v_i) = i$ for all i ($1 \leq i \leq n$). Since every pair of consecutive positive inte-

© The Author(s), under exclusive license to Springer Nature Switzerland AG 2019
G. Chartrand et al., *How to Label a Graph*, SpringerBriefs
in Mathematics, https://doi.org/10.1007/978-3-030-16863-6_4

Fig. 4.1 A prime graph Q_3

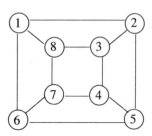

gers are relatively prime (as are 1 and every other positive integer), this is a prime labeling of C_n.   □

By Observation 4.1, $C_3 = K_3$ is therefore a prime graph. However, K_4 is not prime. Regardless of how the vertices of K_4 are labeled with distinct elements of [4], the two vertices labeled 2 and 4 cannot be adjacent. Similarly, K_n is not prime for any integer $n \geq 4$. Indeed, this observation tells us that for each integer $n \geq 4$, graphs of order n and sufficiently large size cannot be prime. That is, the smaller the size of a graph, the more likely it is to be prime.

Since the 4-cycle C_4 is prime and $C_4 = K_{2,2}$, the complete bipartite graph $K_{2,2}$ is prime. In fact, all complete bipartite graphs $K_{2,t}$ where $t \geq 2$ are prime. To see this, it is helpful to make use of a well-known theorem from number theory.

Bertrand's Postulate *For every integer $n \geq 2$, there is at least one prime p such that $n < p < 2n$.*

This theorem is named for Joseph Bertrand, a French mathematician who conjectured in 1845 (in his early 20s) that for every integer $n \geq 4$, there is at least one prime p with $n < p < 2n - 2$ (or for $n \geq 2$, there is at least one prime p with $n < p < 2n$). The Russian mathematician Pafnuty Chebyshev verified this conjecture in 1850, when the resulting theorem became known as *Bertrand's Postulate*. In 1919, the famous Indian mathematician Srinivasa Ramanujan gave a shorter proof of Bertrand's Postulate. In 1932, the 19-year-old Hungarian mathematician Paul Erdős presented (in his first paper) an elementary proof of this fact. Even though Chebyshev initially proved this result, Erdős's proof was universally considered more elegant. What Erdős had accomplished became well known among Hungarian mathematicians and the news of what he had done was often accompanied by the rhyme:

> Chebyshev said it,
> And I say it again.
> There is always a prime
> Between n and $2n$.

Now, we show that the graph $K_{2,t}$ is prime for every integer $t \geq 2$.

Theorem 4.2 *For each integer $t \geq 2$, the complete bipartite graph $K_{2,t}$ is prime.*

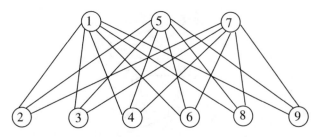

Fig. 4.2 A prime labeling of $K_{3,6}$

Proof Let U and W be the partite sets of $K_{2,t}$ where $|U| = 2$ and $|W| = t$. The labels 1 and the largest prime $p \in [t + 2]$ are assigned to the two vertices of U and the remaining integers in the set $S = [t + 2] - \{1, p\}$ are assigned to the vertices of W. Surely, 1 and every integer in S are relatively prime. Also, p and every integer $k \in S$ with $k < p$ are relatively prime. We claim that p and every integer $k \in S$ with $k > p$ are also relatively prime. If this were not the case, then S must contain an integer $q > p$ such that $p \mid q$ and so $q \geq 2p$. Since $q \in S$, it follows that $2p \in S$. By Bertrand's Postulate, there is a prime p' such that $p < p' < 2p$. But then $p' \in S$ and $p' > p$, which is impossible. Thus, we obtain a prime labeling of $K_{2,t}$. □

Theorem 4.2 then brings up the question of the primality of the complete bipartite graphs $K_{3,t}$ where $t \geq 3$. First, it can be immediately shown that not all such graphs are prime, as $K_{3,3}$ itself is not prime. Not only is the graph $K_{3,3}$ not prime, but for every integer $r \geq 3$, the graph $K_{r,r}$ is not prime.

Theorem 4.3 *For each integer $r \geq 3$, the graph $K_{r,r}$ is not prime.*

Proof Suppose for some integer $r \geq 3$ that $K_{r,r}$ is prime. Let U and W be the partite sets of $K_{r,r}$. Thus, $|U| = |W| = r$. Since half of the integers in $[2r]$ are even, all r of these even integers must be labels of vertices in the same partite set, say U. That is, every vertex of U is labeled with an even integer. Since $r \geq 3$, the vertices labeled 3 and 6 must be in the same partite set. Since the vertex labeled 6 is in U, the vertex labeled 3 is also in U, which is impossible. □

Even though $K_{3,3}$ is not prime, all three graphs $K_{3,4}$, $K_{3,5}$, and $K_{3,6}$ *are* prime. Assigning the labels $1, 5, 7$ to the three vertices in the partite set of size 3 in each of these graphs produces a prime labeling of the graph. A prime labeling of $K_{3,6}$ is shown in Figure 4.2.

Even though $K_{3,4}$, $K_{3,5}$, and $K_{3,6}$ are prime, the graph $K_{3,7}$ is not.

Proposition 4.1 *The complete bipartite graph $K_{3,7}$ is not prime.*

Proof Suppose that there is a prime labeling of $K_{3,7}$ with the integers in the set $[10]$. Let U and W be the partite sets of $K_{3,7}$ where $|U| = 3$ and $|W| = 7$. Since no two

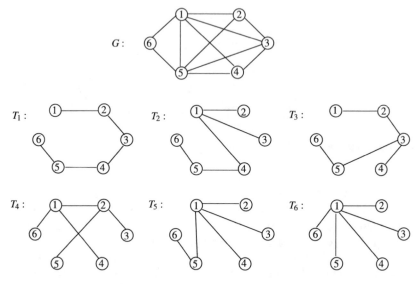

Fig. 4.3 All trees of order 6 are prime

of the integers 2, 4, 6, 8, and 10 are relatively prime, these integers must be assigned to vertices in the same partite set, necessarily to five of the seven vertices of W. However, no two of the integers 3, 6, and 9 are relatively prime and 5 and 10 are not relatively prime. So, 3, 9, and 5 must also be assigned to vertices of W. This leaves only the integers 1 and 7 as the possible labels of the three vertices of U, which is impossible.   □

While $K_{3,7}$ is not prime, it has been shown by M. A. Seoud, A. T. Diab, and E. A. Elsakhawi [43] that $K_{3,t}$ is prime, however, for each integer $t \geq 8$.

The graph G shown in Figure 4.3 is the prime graph of order 6 having maximum size. Therefore, any prime graph of order 6 must be a spanning subgraph of G. As Figure 4.3 also shows, every tree of order 6 is therefore prime, which brings up the question as to which trees are prime (or perhaps not prime).

Trees belonging to certain classes have been shown to be prime, including the following.

1. All paths and stars are prime [17].
2. All trees of order at most 50 are prime [18].
3. All caterpillars with maximum degree at most 5 are prime [50].

In fact, Roger Entringer (see [18, 50]) made the following conjecture in the 1980s.

The Prime Tree Conjecture *Every nontrivial tree is prime.*

In 2011, Penny Haxell, Oleg Pikhurko, and Anusch Taraz [22] proved that all trees having a sufficiently large order are prime.

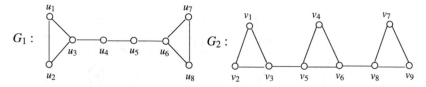

Fig. 4.4 Graphs of order n and size $2\lceil n/2 \rceil + 1$ that are not prime

Theorem 4.4 *There exists a positive integer n' such that for each integer $n > n'$, every tree of order n is prime.*

The Prime Tree Conjecture remains open, however. Since, by Observation 4.1, every cycle is prime and every tree is conjectured to be prime, it is not surprising that M.A. Seoud and M. Z. Youssef [45] made the following conjecture. A *unicyclic graph* is a connected graph with exactly one cycle. Thus, the order and size of a unicyclic graph are the same.

Conjecture 4.1 Every unicyclic graph is prime.

From what we've seen then, every connected graph of order n whose size is either $n - 1$ or n is either prime or is conjectured to be prime. In [22], Haxell, Pikhurko, and Taraz observed that for each integer $n \geq 8$, there exists a graph of order n and size $2\lceil n/2 \rceil + 1$ that is not prime. The graphs G_1 and G_2 of Figure 4.4 have this property. For example, if there were a prime labeling of G_1 with integers in the set [8], then at most one vertex among u_1, u_2, u_3 can have an even label, at most one of u_4, u_5 can have an even label and at most one of u_6, u_7, u_8 can have an even label, which is impossible.

In the same paper [22], the following was verified, thereby establishing the truth of an earlier conjecture by S. N. Rao [37].

Theorem 4.5 *There exists a positive integer n' such that for each integer $n > n'$, the minimum size of a non-prime graph of order n is $2\lceil n/2 \rceil + 1$.*

Since the size of a unicyclic graph equals its order, it follows by Theorem 4.5 that Conjecture 4.1 is true for unicyclic graphs of sufficiently large even order.

If a graph G of order n is not prime, then, by definition, there is no prime labeling of G with labels in the set $[n]$. However, there is always a sufficiently large integer $k \geq n$ and a vertex labeling of G using n distinct integers in the set $[k]$ such that the labels of every two adjacent vertices of G are relatively prime. Such a labeling is often called a *coprime labeling* of G. The smallest such integer k that accomplishes this is called the *coprime index* of G, denoted by $\mathrm{cp}(G)$. For example, if we assign the labels $1, 3, 5$ to one partite set of $K_{3,3}$ and the labels $2, 4, 7$ to the other partite set, then we obtain a coprime labeling of $K_{3,3}$ using six integers in the set [7]. Hence, $\mathrm{cp}(K_{3,3}) = 7$. Similarly, we can assign the labels $1, 3, 5, 9$ to one partite set of $K_{4,4}$ and the labels $2, 4, 7, 8$ to the other partite set to obtain a coprime labeling of $K_{4,4}$ using eight integers in the set [9]. Thus, $\mathrm{cp}(K_{4,4}) = 9$. The values of $\mathrm{cp}(K_{r,r})$

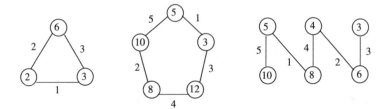

Fig. 4.5 gcd-rainbow labelings of K_3, C_5, and P_6

$(3 \leq r \leq 13)$ were determined in [46]. The values of $\mathrm{cp}(K_{r,r})$ have been determined for even larger values of r.

r	3	4	5	6	7	8	9	10	11	12	13
$\mathrm{cp}(K_{r,r})$	7	9	11	15	17	21	23	27	29	32	27

There is a concept related to the coprime index. Suppose that G is a graph of order n with coprime index k. Then there exists a vertex labeling ℓ of G with n distinct elements of the set $[k]$ where every edge uv of G is assigned the color $\gcd(\ell(u), \ell(v)) = 1$, that is, there exists a monochromatic edge coloring of G, where every edge of G is colored 1. Another well-studied edge coloring of a graph is that of a *rainbow coloring*, where no two edges of the graph are colored the same. For a nonempty graph G of order n and an integer $k \geq n$, a gcd-*rainbow labeling* of G is a vertex labeling ℓ with n distinct integers in the set $[2, k]$ such that each edge uv of G is assigned the color $\gcd(\ell(u), \ell(v))$ and all edges of G have distinct colors. The gcd-*rainbow index* $\mathrm{gr}(G)$ is the smallest integer $k \geq n+1$ such that G has a gcd-rainbow labeling with n elements of $[2, k]$. For example, $\mathrm{gr}(K_3) = 6$, $\mathrm{gr}(C_5) = 12$, and $\mathrm{gr}(P_6) = 10$. See Figure 4.5.

That the gcd-rainbow index exists for every nonempty graph is established next.

Theorem 4.6 *Every nonempty graph has a* gcd-*rainbow labeling.*

Proof   Let $G = K_n$ for $n \geq 2$ with $V(G) = \{v_1, v_2, \ldots, v_n\}$ and let $S = \{p_1, p_2, \ldots, p_n\}$ be the set consisting of the first n primes. Then there are $\binom{n}{n-1} = n$ distinct $(n-1)$-element subsets $S_1, S_2, \ldots, S_n$ of S. For $i = 1, 2, \ldots, n$, the vertex v_i is assigned the label which is the product of the elements of S_i. Therefore, the vertices of G have distinct labels. For vertices u and v of G, the edge uv is assigned the color which is the greatest common divisor of the labels of u and v. Thus, the colors of the edges of G are the products of the elements in the $\binom{n}{n-2} = \binom{n}{2}$ distinct $(n-2)$-element subsets of S. Thus, this is a gcd-rainbow labeling of G. Since any graph H of order n is a spanning subgraph of G, the restriction of a gcd-rainbow labeling of G to H is a gcd-rainbow labeling of H. Therefore, every nonempty graph has a gcd-rainbow labeling.   □

The proof of Theorem 4.6 shows for a graph G of order n that $\mathrm{gr}(G) \leq \prod_{i=2}^{n} p_i$, where $p_1, p_2, \ldots, p_n$ are the first n primes. In particular, $\mathrm{gr}(K_3) \leq 15$. However, we have already seen that $\mathrm{gr}(K_3) = 6$.

For all three graphs shown in Figure 4.5, the edge colors are the elements of the set $[m]$, where m is the size of the graph. Among the questions involving this concept is therefore the following.

Problem 4.1 For which graphs G of size m is there a gcd-rainbow labeling of G whose set of edge colors is $[m]$?

4.2 Multi-prime Labelings

We now describe another vertex labeling of graphs where, once again, adjacent vertices are labeled with relatively prime integers. A vertex labeling f of a nonempty connected graph G is a *multi-prime labeling* if $f : V(G) \to [2, \infty)$, where $uv \in E(G)$ if and only if $\gcd(f(u), f(v)) = 1$. There are some major differences between a multi-prime labeling and a prime (or coprime) labeling. First, rather than the vertex labels coming from a set $[n]$ (or $[k]$), they come from the set $[2, \infty)$. Second, a multi-prime labeling is not required to be injective. Third, adjacent vertices are the *only* pairs of vertices whose labels are relatively prime. Thus, in a multi-prime labeling, 1 is not the label of any vertex and so every label is divisible by at least one prime. Furthermore, if u and v are two nonadjacent vertices, then their labels are not relatively prime and so there is at least one prime that divides both labels. There is no requirement in a multi-prime labeling of a graph that distinct vertices must be assigned distinct labels. If two vertices are labeled the same, however, then these vertices cannot be adjacent.

A multi-prime labeling of the complete graph K_n of order $n \geq 3$ can be obtained by assigning distinct primes to distinct vertices. Thus, the vertex labeling of C_3 shown in Figure 4.6 is a multi-prime labeling. Figure 4.6 also shows a multi-prime labeling of $C_4 = K_{2,2}$. In fact, a multi-prime labeling of a complete multipartite graph G can be obtained by assigning the same prime to two vertices of G if and only if these two vertices belong to the same partite set.

Next, we consider the 5-cycle C_5. A multi-prime labeling of C_5 is shown in Figure 4.7. In this labeling, each label is divisible by at least one of the five primes 2, 3, 5, 7, 11 (actually by exactly two of these primes). One might ask if there is a multi-prime labeling f of C_5 where fewer primes are used. First, we observe that

Fig. 4.6 Multi-prime
labelings of C_3 and C_4

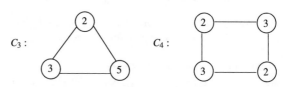

Fig. 4.7 A multi-prime
labeling of C_5

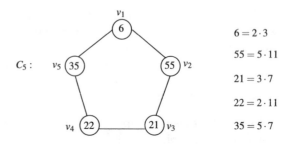

no label in a multi-prime labeling can be a prime. Suppose, say, that $f(v_1) = p$ is a prime. In this case, since $v_1 v_3$ and $v_1 v_4$ are not edges of C_5, it follows that $p \mid f(v_3)$ and $p \mid f(v_4)$, which is impossible since v_3 and v_4 are adjacent, which requires $f(v_3)$ and $f(v_4)$ to be relatively prime. Thus, every label must be divisible by at least two distinct primes and the labels of two adjacent vertices must be divisible by at least one of four distinct primes, say 2, 3, 5, 7. Suppose then that each vertex label of C_5 is divisible by one of the four primes 2, 3, 5, 7. No vertex label can be divisible by three of these primes, for otherwise, the label of each neighbor of this vertex is divisible by a single prime. Therefore, each vertex label is divisible by exactly two primes, say $f(v_1) = 2 \cdot 3$ and $f(v_2) = 5 \cdot 7$. We may assume that $2 \mid f(v_3)$ and $3 \mid f(v_4)$. Also, we may assume that $5 \mid f(v_4)$ and $7 \mid f(v_5)$. However then, $2 \mid f(v_5)$, which is impossible since v_1 and v_5 are adjacent and $2 \mid f(v_1)$. Therefore, every multi-prime labeling of C_5 requires the use of at least five primes. Thus, five primes is the smallest number of primes that can be used in a multi-prime labeling of C_5.

This example brings up a parameter associated with multi-prime labelings of graphs. For a nonempty connected graph G, the *multi-prime index* $\rho(G)$ is the minimum number of primes p in a multi-prime labeling f of G such that $p \mid f(x)$ for at least one vertex x of G. Hence, $\rho(G) \geq 2$ for every nonempty connected graph G. Thus, $\rho(K_n) = n$ and, as we have observed, $\rho(C_5) = 5$. The following observation is often useful.

Observation 4.7 *If H is an induced subgraph of a graph G, then $\rho(H) \leq \rho(G)$.*

For a given nontrivial connected graph G, there is a class of graphs associated with G, all of which have the same multi-prime index as G. To see this, let $V(G) = \{v_1, v_2, \ldots, v_n\}$ and let H be the graph obtained from G by replacing each vertex v_i $(1 \leq i \leq n)$ of G with the empty graph $\overline{K}_{q_i}$ of order q_i. Hence, the vertex set of H is $\cup_{i=1}^{n} V(\overline{K}_{q_i})$ and two vertices u and w of H are adjacent in H if $u \in V(\overline{K}_{q_i})$ and $w \in V(\overline{K}_{q_j})$ where $v_i v_j \in E(G)$. The graph H is referred to as the *composition graph* of G and $\overline{K}_{q_1}, \overline{K}_{q_2}, \ldots, \overline{K}_{q_n}$ and is often denoted by $G[\overline{K}_{q_1}, \overline{K}_{q_2}, \ldots, \overline{K}_{q_n}]$.

Proposition 4.2 *For a nontrivial connected graph G with $V(G) = \{v_1, v_2, \ldots, v_n\}$, let H be the composition graph of G and $\overline{K}_{q_1}, \overline{K}_{q_2}, \ldots, \overline{K}_{q_n}$. Then $\rho(H) = \rho(G)$.*

Proof Suppose that $\rho(G) = k$. For each integer i with $1 \leq i \leq n$, let $w_i \in V(\overline{K}_{q_i})$ and let $S = \{w_1, w_2, \ldots, w_n\}$. Since $H[S] \cong G$ and $H[S]$ is an induced subgraph of H, it follows by Observation 4.7 that $k = \rho(G) = \rho(H[S]) \leq \rho(H)$. Next, let there be given a multi-prime labeling $f : V(G) \to [2, \infty)$ of G. Define the labeling $g : V(H) \to [2, \infty)$ by $g(u) = f(v_i)$ if $u \in V(\overline{K}_{q_i})$ where $1 \leq i \leq n$. Since g is a multi-prime labeling of H, it follows that $\rho(H) \leq k = \rho(G)$. Therefore, $\rho(H) = \rho(G) = k$. $\qquad\qquad\square$

Since a complete k-partite graph G is the composition graph of K_k and k empty graphs, the following is an immediate consequence of Proposition 4.2.

Corollary 4.1 *If G is a complete k-partite graph, $k \geq 2$, then $\rho(G) = \rho(K_k) = k$.*

As noted earlier, in a multi-prime labeling of a graph, the vertex labels are not required to be distinct. However, if two labels are the same, such as can occur with vertices belonging to the same partite set of a complete multipartite graph, primes can be raised to higher powers to produce a multi-prime labeling with distinct labels without increasing the number of primes used. For example, if a multi-prime labeling of a complete multipartite graph G assigns a prime p to every vertex in a partite set of G and the partite set consists of $r \geq 2$ vertices, then $p, p^2, \ldots, p^r$ may be assigned to these vertices to produce distinct labels. This, however, has no effect on the value of $\rho(G)$.

Unlike the situation for prime labelings, where a graph may or may not have such a labeling, every nontrivial connected graph has a multi-prime labeling. Let $\mathscr{P} = \{p_1, p_2, p_3, \ldots\}$ be the set of distinct primes, where we can assume that $p_1 = 2 < p_2 = 3 < p_3 = 5 < \ldots$ and so on.

Theorem 4.8 *Every nontrivial connected graph has a multi-prime labeling.*

Proof We proceed by induction on the order n of a connected graph. The result is immediate for small values of n, say $n = 2, 3, 4$. Assume that the statement is true for all connected graphs of order n for an integer $n \geq 4$. Let G be a connected graph of order $n + 1$, let v be a non-cut-vertex of G, and let $G' = G - v$. Since G' is a connected graph of order n, it follows by the induction hypothesis that G' has a multi-prime labeling. Let such a labeling f' of G' be given using k primes, say $p_1, p_2, \ldots, p_k$. Suppose that $V(G') = \{u_1, u_2, \ldots, u_n\}$, where $N(v) = \{u_1, u_2, \ldots, u_r\}$, $1 \leq r \leq n$. Define a vertex labeling f of G by

$$f(x) = \begin{cases} p_{i+k} \, f'(u_i) & \text{if } x = u_i \text{ for } 1 \leq i \leq n \\ p_{n+k+1} \prod_{i=k+r+1}^{k+n} p_i & \text{if } x = v. \end{cases}$$

Since $xy \in E(G)$ if and only if $\gcd(f(x), f(y)) = 1$, it follows that f is a multi-prime labeling of G. $\qquad\qquad\square$

4.3 Subset Labelings

The multi-prime labelings encountered in the preceding section can be looked at in another way. First, let us review the concept of a multi-prime labeling f of a graph G. Once again, let $2 = p_1, 3 = p_2, 5 = p_3, p_4, p_5, \ldots$ be the primes. Then for two vertices u and v of G, it follows that u and v are assigned labels $f(u) = p_{i_1} p_{i_2} \cdots p_{i_k}$ and $f(v) = p_{j_1} p_{j_2} \cdots p_{j_\ell}$, where $p_{i_1}, p_{i_2}, \ldots, p_{i_k}$ are distinct primes and $p_{j_1}, p_{j_2}, \ldots, p_{j_\ell}$ are distinct primes. If u and v are adjacent, then all $k + \ell$ primes are distinct and so $\{i_1, i_2, \ldots, i_k\} \cap \{j_1, j_2, \ldots, j_\ell\} = \emptyset$; while if u and v are not adjacent, then at least one of the primes $p_{i_1}, p_{i_2}, \ldots, p_{i_k}$ is the same as one of the primes $p_{j_1}, p_{j_2}, \ldots, p_{j_\ell}$, that is, $\{i_1, i_2, \ldots, i_k\} \cap \{j_1, j_2, \ldots, j_\ell\} \neq \emptyset$. It therefore follows that the multi-prime labeling f can be considered as a function $f : V(G) \to \mathscr{P}^*([r])$ for some integer $r \geq 2$, where $\mathscr{P}^*([r])$ is the set of nonempty subsets of $[r]$, such that $f(u) \cap f(v) = \emptyset$ if and only if u and v are adjacent vertices of G. Consequently, rather than assigning the integer $f(u) = p_{i_1} p_{i_2} \cdots p_{i_k}$ to a vertex u of G, we can assign the subset $f(u) = \{i_1, i_2, \ldots, i_k\}$ of $[r]$ to u. Furthermore, the minimum positive integer r for which such a function exists is the number $\rho(G)$. When expressed in this manner, we refer to such a vertex labeling of a graph G as a *subset labeling* of G. Therefore, the concepts of multi-prime labeling and subset labeling are essentially the same concept and $\rho(G)$ is the same parameter in each case.

For example, three (essentially equivalent) multi-prime labelings and the corresponding subset labeling of a graph G of order 5 are shown in Figure 4.8. In the drawing of a graph, we write $\{a, b\}$ as ab for simplicity. The first labeling uses the specific primes 2, 3, 5, the next two labelings refer to these three primes as p_1, p_2, p_3, and the fourth one describes the subscripts of these three primes as subsets of $[3]$ (where, as indicated above, we write $\{1, 2\}$ as 12). While the second labeling assigns the label $p_1 p_2$ to two vertices of G, the third labeling assigns distinct labels to distinct vertices of G. Since G contains K_3 as an induced subgraph, it follows by Observation 4.7 that $\rho(G) = 3$.

As another example, consider the Petersen graph P of order 10. Figure 4.9 shows a subset labeling of P using labels in $\mathscr{P}^*([5])$. Thus, $\rho(P) \leq 5$. Since P contains the 5-cycle as an induced subgraph, it follows that $\rho(P) \geq \rho(C_5) = 5$ by

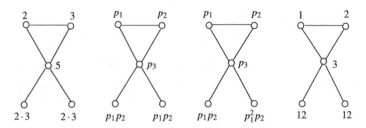

Fig. 4.8 A graph G with $\rho(G) = 3$

Fig. 4.9 The Petersen graph
P has $\rho(P) = 5$

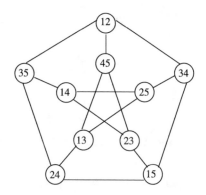

Observation 4.7. Hence, $\rho(P) = 5$. An interesting feature of the subset labeling of P shown in Figure 4.9 is that this labeling uses all ten distinct 2-element subsets of $[5]$ for the ten vertices of P. Actually, this vertex labeling of the Petersen graph is a familiar one. It is well known that if we define a graph $G = (V, E)$ such that the vertex set V of G is the set of all $\binom{5}{2} = 10$ distinct 2-element subsets of $[5]$ and the edge set E is the set of all pairs of disjoint 2-element subsets of V, then the resulting graph G is isomorphic to the Petersen graph.

The Petersen graph is a member of a special class of graphs. For positive integers k and n with $n > 2k$, the *Kneser graph* $KG_{n,k}$ is that graph of order $\binom{n}{k}$ whose vertices are labeled with distinct elements of the set $\mathscr{P}_k([n])$ of k-element subsets of the set $[n]$ where two vertices are adjacent if and only if they have disjoint labels. The graph $KG_{n,k}$ is therefore $\binom{n-k}{k}$-regular. In particular, $KG_{n,1}$ is the complete graph K_n and $KG_{5,2}$ is the Petersen graph. The Kneser graphs are named for Martin Kneser, who, while investigating partitions of sets, made the following conjecture in 1955 [27] (stated in terms of graphs).

Kneser's Conjecture For positive integers k and n with $n > 2k$, there exists no proper $(n - 2k + 1)$-coloring of the Kneser graph $KG_{n,k}$.

Lovász [31] verified this conjecture in 1978 when he proved the following result.

Theorem 4.9 *For every two positive integers k and n with $n > 2k$,*

$$\chi(KG_{n,k}) = n - 2k + 2.$$

For integers $n \geq 2$, the Kneser graphs $O_n = KG_{2n-1,n-1}$, called *odd graphs*, have been of special interest. The graph O_2 is K_3, the graph O_3 is the Petersen graph, and O_4 is a 4-regular graph of order 35. By Theorem 4.9, every odd graph has chromatic number 3.

We now investigate the multi-prime index $\rho(G)$ for some specific graphs G. In each case, the labelings involved are subset labelings. We begin with paths. First, we evaluate $\rho(G)$ when G is a small path.

Proposition 4.3 *For $3 \leq n \leq 6$, $\rho(P_n) = n - 1$.*

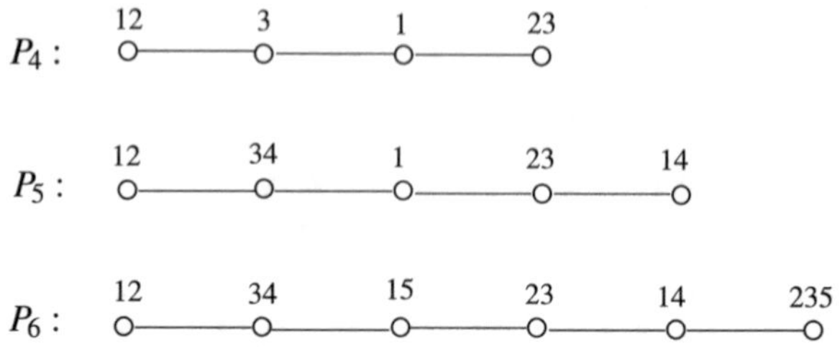

$P_4:$ 12 —— 3 —— 1 —— 23

$P_5:$ 12 —— 34 —— 1 —— 23 —— 14

$P_6:$ 12 —— 34 —— 15 —— 23 —— 14 —— 235

Fig. 4.10 Subset labelings of P_4, P_5, and P_6

Proof Clearly, $\rho(P_3) = 2$. The subset labelings of P_n in Figure 4.10 show that $\rho(P_n) \leq n - 1$ for $4 \leq n \leq 6$. Again, in the figure, the sets $\{a\}$, $\{a, b\}$, $\{a, b, c\}$ are denoted by a, b, abc, respectively. We only show that $\rho(P_6) = 5$. If f is a subset labeling of $P_6 = (v_1, v_2, \ldots, v_6)$ using subsets in [4], then there is no vertex v of P_6 for which $|f(v)| = 1$, for suppose that $f(v_j) = \{1\}$ where $1 \leq j \leq 6$. Then $1 \in f(u)$ if u is not a neighbor of v_j. Since at least two such vertices are adjacent, this is impossible. There is no vertex v of P_6 with $|f(v)| = 3$ either, for this would imply that the label of each neighbor of v is a singleton. Consequently, the label of each vertex of P_6 is a 2-element subset of [4]. However, this implies that $f(v_i) = \{1, 2\}$, say, where i is odd and $f(v_i) = \{3, 4\}$, where i is even. Since v_1 and v_4 are not adjacent and $f(v_1) \cap f(v_4) = \emptyset$, this is a contradiction. Hence, $\rho(P_6) = 5$.   □

In fact, 6 is the largest value of n for which $\rho(P_n) = n - 1$. In order to show this, we first present the following result.

Theorem 4.10 *For an integer $n \geq 7$, let $P_n = (v_1, v_2, \ldots, v_n)$ and let k be an integer with $k \geq \rho(P_n)$. Suppose that there is a subset labeling $f_n : V(P_n) \to \mathscr{P}^*([k])$ for which there are two distinct integers a and b such that $a \in f_n(v_i)$ for each even integer i and $b \in f_n(v_i)$ for each odd integer i where $1 \leq i \leq n$. Then there is a subset labeling $f_{n+1} : V(P_{n+1}) \to \mathscr{P}^*([k + 1])$ of $P_{n+1} = (v_1, v_2, \ldots, v_{n+1})$ such that $a \in f_{n+1}(v_i)$ for each even integer i and $b \in f_{n+1}(v_i)$ for each odd integer i where $1 \leq i \leq n + 1$.*

Proof For $P_{n+1} = (v_1, v_2, \ldots, v_n, v_{n+1})$, define the labeling $f_{n+1} : V(P_{n+1}) \to \mathscr{P}^*([k + 1])$ from the labeling f_n of P_n as follows:

★   If n is odd, then

$$f_{n+1}(v_i) = \begin{cases} f_n(v_i) & \text{if } i \text{ is even for } 2 \leq i \leq n - 1 \text{ or } i = n \\ f_n(v_i) \cup \{k + 1\} & \text{if } i \text{ is odd for } 1 \leq i \leq n - 2 \\ \{a, k + 1\} & \text{if } i = n + 1. \end{cases}$$

★   If n is even, then

$$f_{n+1}(v_i) = \begin{cases} f_n(v_i) & \text{if } i \text{ is odd for } 1 \leq i \leq n - 2 \text{ or } i = n \\ f_n(v_i) \cup \{k + 1\} & \text{if } i \text{ is even for } 2 \leq i \leq n - 2 \\ \{b, k + 1\} & \text{if } i = n + 1. \end{cases}$$

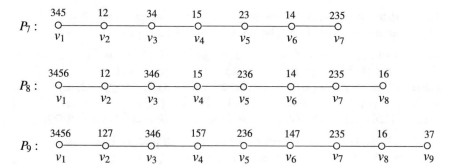

Fig. 4.11 Subset labelings of P_n for $n = 7, 8, 9$

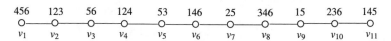

Fig. 4.12 A subset labeling of P_{11}

Then f_{n+1} is a subset labeling of P_{n+1} with the desired properties.   □

Figure 4.11 shows a subset labeling of P_7, which implies that $\rho(P_7) \leq 5$. However, since $\rho(P_6) = 5$, it follows that $\rho(P_7) = 5$. For $n = 7$ and $n = 8$, the subset labelings f_{n+1} of P_8 and P_9 in Figure 4.11 are those defined in the proof of Theorem 4.10. As before, we write $\{a, b, c, \ldots\}$ as $abc \cdots$.

The following is a consequence of Theorem 4.10.

Corollary 4.2 *For each integer $n \geq 7$, $\rho(P_n) \leq n - 2$.*

Proof The subset labeling $f_7 : V(P_7) \rightarrow \mathscr{P}^*([5])$ of P_7 shown in Figure 4.11 has the desired property that $1 \in f_7(v_i)$ for each even integer i and $3 \in f_n(v_i)$ for each odd integer i where $1 \leq i \leq 7$. Repeatedly applying Theorem 4.10, we obtain, for an integer $n \geq 7$, a subset labeling $f_n : V(P_n) \rightarrow \mathscr{P}^*([n - 2])$ of $P_n = (v_1, v_2, \ldots, v_n)$ such that $1 \in f_n(v_i)$ for each even integer i and $3 \in f_n(v_i)$ for each odd integer i, where $1 \leq i \leq n$. Therefore, $\rho(P_n) \leq n - 2$ for $n \geq 7$.   □

By Corollary 4.2, $\rho(P_8) \leq 6$; in fact, $\rho(P_8) = 6$. However, 8 is the largest value of n for which $\rho(P_n) = n - 2$.

Proposition 4.4 *For $8 \leq n \leq 11$, $\rho(P_n) = 6$.*

Proof Figure 4.12 shows a subset labeling $f : V(P_{11}) \rightarrow \mathscr{P}^*([6])$ of P_{11} and so $\rho(P_{11}) \leq 6$. It then follows by Observation 4.7 that $6 = \rho(P_8) \leq \rho(P_n) \leq 6$ for $8 \leq n \leq 11$. Hence, it remains to show that $\rho(P_8) \geq 6$.

Assume, to the contrary, that $\rho(P_8) < 6$. Since $\rho(P_7) = 5$, it follows that $\rho(P_8) = 5$ and so there is a subset labeling $g : V(P_8) \rightarrow \mathscr{P}^*([5])$ of P_8. Observe that

(1) no vertex of P_8 can be labeled with a single integer,

(2) no vertex of P_8 can be labeled with four or more integers, and

(3) no interior vertex of P_8 can be labeled with three integers.

Let $P_8 = (v_1, v_2, \ldots, v_8)$. If $|g(v_1)| = 3$, say $g(v_1) = \{1, 2, 3\}$, then we may assume that $g(v_2) = \{4, 5\}$, $g(v_3) = \{1, 2\}$, $g(v_4) = \{3, 4\}$, $g(v_5) = \{1, 5\}$, $g(v_6) = \{2, 4\}$. However then, $g(v_7) = \{1, 3, 5\}$, which contradicts (3). If $|g(v_1)| = 2$, say $g(v_1) = \{1, 2\}$, then we may assume that $g(v_2) = \{3, 4\}$, $g(v_3) = \{1, 5\}$, $g(v_4) = \{2, 3\}$, $g(v_5) = \{1, 4\}$. However then, $g(v_6) = \{2, 3, 5\}$, which again contradicts (3).   □

Since P_n is an induced subgraph of P_{n+1} for each positive integer n, it follows by Observation 4.7 that $\rho(P_n) \leq \rho(P_{n+1})$ and so $\{\rho(P_n)\}, n \geq 2$, is a nondecreasing sequence of positive integers. First, we establish another property of this sequence.

Theorem 4.11 *For an integer $n \geq 3$,*

$$\rho(P_n) \leq \rho(P_{n+1}) \leq \rho(P_n) + 1.$$

Proof That $\rho(P_n) \leq \rho(P_{n+1})$ follows from Observation 4.7, as noted above. We now show that $\rho(P_{n+1}) \leq \rho(P_n) + 1$. Let $P_n = (v_1, v_2, \ldots, v_n)$ and $P_{n+1} = (v_1, v_2, \ldots, v_n, v_{n+1})$. Suppose that $\rho(P_n) = r$. Let $g : V(P_n) \to \mathscr{P}^*([r])$ be a multi-prime labeling of P_n. We now define a subset labeling $f : V(P_{n+1}) \to \mathscr{P}^*([r + 1])$ by

$$f(v_i) = \begin{cases} g(v_i) & \text{if } 1 \leq i \leq n \text{ with } i \neq n - 2 \\ g(v_i) \cup \{r + 1\} & \text{if } i = n - 2 \\ [r + 1] - g(v_n) & \text{if } i = n + 1. \end{cases}$$

Therefore, $f(v_i) \cap f(v_{i+1}) = \emptyset$ for $1 \leq i \leq n$ and $f(v_i) \cap f(v_j) \neq \emptyset$ for $1 \leq i, j \leq n$ and $|i - j| \geq 2$. It remains to show that $f(v_{n+1}) \cap f(v_i) \neq \emptyset$ for $1 \leq i \leq n - 1$. Since $f(v_n) \cap f(v_{n-1}) = \emptyset$ and $f(v_{n+1}) = [r + 1] - g(v_n)$, there exists an integer $j \in [r]$ such that $j \in f(v_{n+1}) \cap f(v_{n-1})$. Thus, $f(v_{n+1}) \cap f(v_{n-1}) \neq \emptyset$. Since $r + 1 \in f(v_{n+1}) \cap f(v_{n-2})$, it follows that $f(v_{n+1}) \cap f(v_{n-2}) \neq \emptyset$. For each integer i with $1 \leq i \leq n - 3$, there exists an integer $\ell_i \in f(v_i)$ such that $\ell_i \in f(v_{n-1})$. Since $f(v_{n-1}) \cap f(v_n) = \emptyset$, it follows that $\ell_i \in f(v_{n+1})$. Thus, $f(v_{n+1}) \cap f(v_i) \neq \emptyset$ for $1 \leq i \leq n - 3$. Hence, f is a multi-prime labeling of P_{n+1} and so $\rho(P_{n+1}) \leq r + 1 = \rho(P_n) + 1$.   □

The following result establishes another property of the sequence $\{\rho(P_n)\}$.

Theorem 4.12 $\lim_{n \to \infty} \rho(P_n) = \infty$.

Proof Assume, to the contrary, that $\lim_{n \to \infty} \rho(P_n) \neq \infty$. By Theorem 4.11, there exist positive integers N and r such that $\rho(P_n) = r$ for every $n \geq N$. Hence, there is a sufficiently large positive integer n such that for the path $P_n = (v_1, v_2, \ldots, v_n)$ and a multi-prime labeling $f : V(P_n) \to \mathscr{P}^*([r])$, there are distinct vertices v_i, v_j, v_k of P_n with $i < j < k$ for which $f(v_i) = f(v_j) = f(v_k) = S \subseteq [r]$. Thus, $j - i \geq 2$ and $k - j \geq 2$. Since $f(v_i) \cap f(v_{k-1}) \neq \emptyset$, there exists $\ell \in f(v_i) \cap f(v_{k-1})$. Thus, $\ell \in S$. However then, $\ell \in f(v_{k-1}) \cap f(v_k)$ and so $f(v_{k-1}) \cap f(v_k) \neq \emptyset$, a contradiction.   □

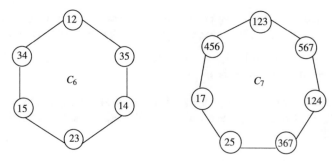

Fig. 4.13 Subset labelings of C_6 and C_7

The following is a consequence of Theorems 4.11 and 4.12.

Corollary 4.3 *For each integer $r \geq 3$, there exists an integer n_r such that*

$$\rho(P_{n_r}) = r.$$

We now turn our attention to cycles. We have seen that $\rho(C_3) = 3$, $\rho(C_4) = 2$, and $\rho(C_5) = 5$.

Proposition 4.5 $\rho(C_6) = 5$ *and* $\rho(C_7) = 7$.

Proof The subset labelings of C_6 and C_7 in Figure 4.13 show that $\rho(C_6) \leq 5$ and $\rho(C_7) \leq 7$. We only show that $\rho(C_7) = 7$. Let $C_7 = (v_1, v_2, \ldots, v_7, v_1)$.

Assume, to the contrary, that there is a subset labeling $f : V(C_7) \to \mathscr{P}^*([6])$ of C_7. We begin with some observations.

(1) There is no vertex v of C_7 with $|f(v)| = 1$. Suppose that $f(v_1) = \{1\}$. Since v_3 and v_4 are not adjacent to v_1, it follows that $1 \in f(v_3)$ and $1 \in f(v_4)$. However, v_3 and v_4 are adjacent and $f(v_3) \cap f(v_4) \neq \emptyset$, a contradiction.
(2) There is no vertex v of C_7 with $|f(v)| = 5$. Suppose that $f(v_1) = \{1, 2, 3, 4, 5\}$. Thus, $f(v_2) = 6$, which is impossible by (1).
(3) There is no vertex v of C_7 with $|f(v)| = 4$. Suppose that $f(v_1) = \{1, 2, 3, 4\}$. Since $f(v_1) \cap f(v_2) = \emptyset$ and $f(v_1) \cap f(v_7) = \emptyset$, it follows by (1) that $f(v_2) = f(v_7) = \{5, 6\}$. Since $f(v_2) \cap f(v_3) = \emptyset$, it follows that $f(v_3) \subseteq \{1, 2, 3, 4\}$ and so $f(v_3) \cap f(v_7) = \emptyset$. However, v_3 and v_7 are not adjacent. This is a contradiction.
(4) There do not exist adjacent vertices u and v of C_7 with $|f(u)| = |f(v)| = 3$. Suppose that $f(v_1) = \{1, 2, 3\}$ and $f(v_2) = \{4, 5, 6\}$. Since $f(v_2) \cap f(v_3) = \emptyset$, it follows that $f(v_3) \subseteq \{1, 2, 3\}$ and $f(v_7) \subseteq \{4, 5, 6\}$ and so $f(v_3) \cap f(v_7) = \emptyset$, again a contradiction.
(5) There do not exist vertices u and v of C_7 having a common neighbor such that $|f(u)| = |f(v)| = 2$. Suppose that $|f(v_1)| = |f(v_3)| = 2$ and $f(v_1) = \{1, 2\}$. We may assume that $1 \in f(v_3)$, $2 \in f(v_4)$, $1 \in f(v_5)$, and $2 \in f(v_6)$. Since $|f(v_3)| = 2$ and $2 \in f(v_4)$, we may further assume that $f(v_3) = \{1, 3\}$. Because

v_3 is adjacent to neither v_6 nor v_7 and $1 \in f(v_5) \cap f(v_1)$, it follows that $3 \in f(v_6)$ and $3 \in f(v_7)$. However, v_6 and v_7 are adjacent, which is impossible.

By (1)–(3), $|f(v)| = 2$ or $|f(v)| = 3$ for every vertex v of C_7. However, (4) and (5) imply that no such labeling exists.   □

For each integer $n \geq 3$, the path P_{n-2} is an induced subgraph of the n-cycle C_n. Thus, $\rho(P_{n-2}) \leq \rho(C_n)$ by Observation 4.7. Therefore, the following is a consequence of Theorem 4.12.

Proposition 4.6 $\lim_{n\to\infty} \rho(C_n) = \infty.$

While $\lim_{n\to\infty} \rho(C_n) = \infty$ by Proposition 4.6, it follows by Proposition 4.5 that there is no result for cycles that corresponds to Theorem 4.11 for paths.

The results mentioned above suggest the following questions.

Problem 4.2 What is $\rho(P_n)$ for each integer $n \geq 2$?

Problem 4.3 What is $\rho(C_n)$ for each integer $n \geq 3$?

We have now seen examples of graphs whose multi-prime index equals its chromatic number and others whose multi-prime index exceeds its chromatic number. As we now show, there is no other possibility.

Theorem 4.13 *If G is a nontrivial connected graph, then $\rho(G) \geq \chi(G)$.*

Proof Let $\rho(G) = r \geq 2$ and let $f : V(G) \to \mathscr{P}^*([r])$ be a multi-prime labeling of G. Define the vertex coloring $c : V(G) \to [r]$ by $c(x) = \min\{i \in [r] : i \in f(x)\}$. Let u and v be two adjacent vertices of G. Since $f(u) \cap f(v) = \emptyset$, it follows that $c(u) \neq c(v)$. Thus, c is a proper coloring of G using at most r colors. Therefore, $\chi(G) \leq r = \rho(G)$.   □

We have seen that $\rho(G) \geq 2$ for every nonempty connected graph G. Next, we present a characterization of connected graphs G having $\rho(G) = 2$.

Proposition 4.7 *Let G be a nontrivial connected graph. Then $\rho(G) = 2$ if and only if G is a complete bipartite graph.*

Proof Since $\rho(G) = 2$ for every complete bipartite graph G by Corollary 4.1, it remains to verify the converse. Let G be a nontrivial connected graph with $\rho(G) = 2$. By Theorem 4.13, G is bipartite and no vertex of G can be labeled $\{1, 2\}$. Thus, every vertex of G is labeled $\{1\}$ or $\{2\}$. Let U be the set of vertices of G labeled $\{1\}$ and W be the set of vertices of G labeled $\{2\}$. Therefore, U and W are independent sets. Since $\{1\}$ and $\{2\}$ are disjoint, every vertex of U is adjacent to every vertex of W. That is, G is a complete bipartite graph with partite sets U and W.   □

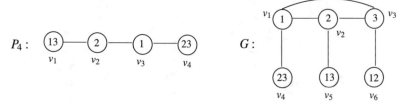

Fig. 4.14 Two graphs with multi-prime index 3

According to Proposition 4.7, if G is a nontrivial connected graph that is not a complete bipartite graph, then $\rho(G) \geq 3$. We saw that $\rho(G) = 3$ for the graph G of Figure 4.8 and that the multi-prime index of every complete 3-partite graph is 3. There are other graphs having multi-prime index 3. We saw in Proposition 4.3 that $\rho(P_4) = 3$; also, $\rho(K_3) = 3$. Furthermore, the graph G of Figure 4.14 also has multi-prime index 3. Multi-prime labelings of P_4 and G are shown in the figure. By Proposition 4.2, a composition graph can be constructed from P_4 and G by replacing each vertex v_i by an empty graph, resulting in a graph having the same multi-prime index as P_4 or K_3.

It follows from Theorem 4.13 that if G is a connected graph with $\rho(G) = 3$, then $\chi(G) = 2$ or $\chi(G) = 3$. Therefore, there are graphs G with $\rho(G) = 3$ for which $\chi(G) = k$ where $k = 2$ or $k = 3$. Similarly, if G is a connected graph with $\rho(G) = 4$, then for each integer $k \in \{2, 3, 4\}$, there exists a connected graph G with $\rho(G) = 4$ and $\chi(G) = k$. This gives rise to the following conjecture.

Conjecture 4.2 For each pair k, ℓ of integers with $2 \leq k \leq \ell$, there exists a connected graph G with $\chi(G) = k$ and $\rho(G) = \ell$.

4.4 2-Prime Labelings

We saw in a multi-prime labeling of a graph G that integers from the set $[2, \infty)$ are assigned to the vertices of G in such a way that two vertices of G are assigned relatively prime integers if and only if they are adjacent. The complement $\overline{G}$ of G therefore has the property that two distinct integers a and b from the set $[2, \infty)$ can be assigned to adjacent vertices of G if and only if a and b are not relatively prime, that is, if there is some prime p such that $p \mid a$ and $p \mid b$. Both the 5-cycle C_5 and the wheel $W_5 = C_4 \vee K_1$ of Figure 4.15 have such a labeling. In both labelings, each vertex label is the product of two distinct primes. These are examples of a particular kind of labeling having a property that is opposite to that of a multi-prime labeling.

An integer is a *2-prime integer* if it is the product of two distinct primes. A *2-prime labeling* of a graph G is an assignment of distinct 2-prime integers to the vertices of G such that two vertices of G are assigned labels that are *not* relatively prime if and only if they are adjacent. Thus, in a 2-prime labeling f of a graph, every two distinct

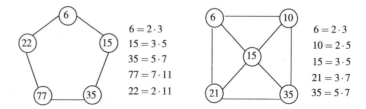

Fig. 4.15 Labeling the vertices of C_5 and $W_5 = C_4 \vee K_1$

Fig. 4.16 A graph with no
2-prime labeling

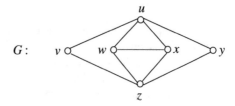

vertices u and v are assigned two distinct 2-prime integers. If u and v are adjacent, then $f(u)$ and $f(v)$ are not relatively prime; while if u and v are not adjacent, then $f(u)$ and $f(v)$ are relatively prime. Therefore, both labelings shown in Figure 4.15 are 2-prime labelings.

Not all graphs can be given 2-prime labelings, however, for consider the graph G of Figure 4.16. Suppose that there is a 2-prime labeling f of G. Then $f(z) = p_1 p_2$, where p_1 and p_2 are distinct primes. Since $vz \in E(G)$, we can assume that $f(v) = p_1 p_3$, where p_3 is a prime distinct from p_1 and p_2. Since $yz \in E(G)$, either $p_1 \mid f(y)$ or $p_2 \mid f(y)$, but not both. Since $f(v) = p_1 p_3$ and $vy \notin E(G)$, it follows that $p_1 \nmid f(y)$; thus, $p_2 \mid f(y)$. Hence, $f(y) = p_2 p_4$ for some prime p_4 distinct from p_1 and p_3. Because $wz \in E(G)$, either $p_1 \mid f(w)$ or $p_2 \mid f(w)$, both of which are impossible since $wv, wy \notin E(G)$. Therefore, no 2-prime labeling of G is possible.

The argument we gave as to why there is no 2-prime labeling of the graph G of Figure 4.16 comes from the fact that G contains $G[\{z, v, w, y\}] = K_{1,3}$ as an induced subgraph and there is no 2-prime labeling of $K_{1,3}$. In a similar manner, one can show that there is no 2-prime labeling of any of the nine graphs shown in Figure 4.17. Thus, no graph containing any of these nine graphs as induced subgraphs has a 2-prime labeling. These nine graphs play a key role in another concept in graph theory.

Let G be a nonempty graph. The *line graph* $L(G)$ of G is that graph whose vertex set consists of the edges of G, where two vertices of $L(G)$ are adjacent if the corresponding edges of G are adjacent. A graph H and its line graph $L(H)$ are shown in Figure 4.18. A graph G is *a line graph* if there exists a graph F such that $G = L(F)$. Thus, the graph $L(H)$ of Figure 4.18 is a line graph.

In 1970, Lowell Beineke [2] characterized the graphs that are line graphs as those graphs containing none of the nine graphs shown in Figure 4.17 as induced subgraphs.

Theorem 4.14 (Beineke's Theorem) *A graph G is a line graph if and only if G does not contain any of the nine graphs of Figure 4.17 as an induced subgraph.*

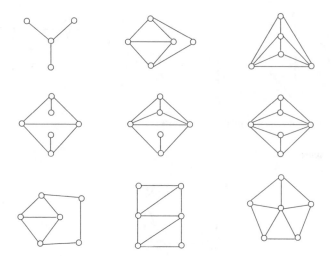

Fig. 4.17 The nine forbidden induced subgraphs of a line graph

In particular, the graph G of Figure 4.16 is not a line graph. Theorem 4.14 allows us to present the following result.

Theorem 4.15 *A graph G possesses a 2-prime labeling if and only if G is a line graph.*

Proof Assume first that G is a graph that possesses a 2-prime labeling. Then each induced subgraph of G has a 2-prime labeling as well. Since none of the nine graphs of Figure 4.17 has a 2-prime labeling, it follows that G contains none of these nine graphs as an induced subgraph. By Theorem 4.14, G is a line graph.

Conversely, assume that G is a line graph. Then there is a graph H such that $G = L(H)$. Suppose that H has order n. Label the n vertices of H with n distinct primes $p_1, p_2, \ldots, p_n$. For an edge e of H joining vertices labeled p_i and p_j, where $1 \leq i, j \leq n$ and $i \neq j$, we assign the 2-prime integer $p_i p_j$ to the edge e. Then the vertex e of $G = L(H)$ is also labeled $p_i p_j$. This produces a vertex labeling of G. Observe that two vertices e' and e'' of G are adjacent in G if and only if the edges e' and e'' of H are incident with a common vertex, labeled p_k say, where then p_k divides the labels of e' and e''. Therefore, this vertex labeling is a 2-prime labeling of G. □

As is the case for multi-prime labelings, 2-prime labelings can also be expressed as a certain type of subset labeling. For example, if G is a graph with a 2-prime labeling in which the primes involved are $p_1, p_2, \ldots, p_n$, then each vertex labeled $p_i p_j$ can be expressed as the 2-element subset $\{i, j\}$ of $[n]$. For example, for the 2-prime labelings of the graphs C_5 and $W_5 = C_4 \vee K_1$ shown in Figure 4.15, we can write $p_1 = 2$, $p_2 = 3$, $p_3 = 5$, $p_4 = 7$, $p_5 = 11$. Expressing the 2-prime integer $p_i p_j$ as the 2-element subset $\{i, j\}$ of [4] or [5], which we write again as ij, we have

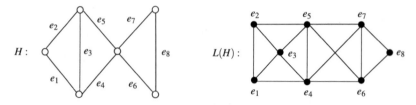

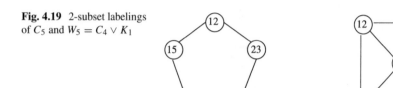

Fig. 4.18 A graph and its line graph

Fig. 4.19 2-subset labelings
of C_5 and $W_5 = C_4 \vee K_1$

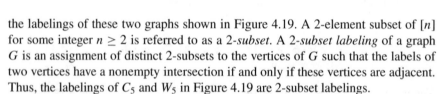

the labelings of these two graphs shown in Figure 4.19. A 2-element subset of $[n]$ for some integer $n \geq 2$ is referred to as a 2-*subset*. A 2-*subset labeling* of a graph G is an assignment of distinct 2-subsets to the vertices of G such that the labels of two vertices have a nonempty intersection if and only if these vertices are adjacent. Thus, the labelings of C_5 and W_5 in Figure 4.19 are 2-subset labelings.

Consequently, the concept of 2-subset labelings of graphs provides us with an alternative way of looking at 2-prime labelings, giving the following corollary of Theorem 4.15.

Corollary 4.4 *A graph G possesses a 2-subset labeling if and only if G is a line graph.*

Therefore, not only are the graphs C_5 and W_5 line graphs (in fact, $C_5 = L(C_5)$ and $W_5 = L(K_{1,1,2})$) but the complement $\overline{P}$ of the Petersen graph P is a line graph. In fact, $\overline{P} = L(K_5)$.

It is useful to describe another concept in graph theory. For a nonempty set A and a collection S of distinct nonempty subsets of A, the *intersection graph* $\Omega(S)$ of S is that graph with vertex set S where two vertices are adjacent if and only if the subsets have a nonempty intersection. For example, let G be a nontrivial connected graph with $V(G) = \{v_1, v_2, \ldots, v_n\}$, where $A = E(G)$, S_i is the set of edges incident with v_i for $1 \leq i \leq n$, and $S = \{S_1, S_2, \ldots, S_n\}$. Then $\Omega(S) \cong G$, that is, every nontrivial connected graph is an intersection graph. Thus, each line graph is an intersection graph where the subsets involved are 2-subsets. The topic of intersection graphs has been discussed in detail in a book by Terry A. McKee and F. R. McMorris [32].

Integers that are k-prime integers and k-prime labelings of graphs for an integer $k \geq 2$ are defined as expected. An integer is a k-*prime integer*, $k \geq 2$, if it is the product of k distinct primes. A k-*prime labeling* of a graph G is an assignment of distinct k-prime integers to the vertices of G such that the labels of two vertices u

Fig. 4.20 A 3-prime
labeling of the Petersen
graph

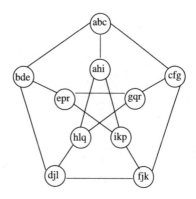

and v of G are *not* relatively prime if and only if u and v are adjacent. For example, a
3-prime labeling of the Petersen graph is shown in Figure 4.20 (where the 15 distinct
primes here are denoted by a, b, c, d, e, f, g, h, i, j, k, l, p, q, r). Since the Petersen
graph contains $K_{1,3}$ as an induced subgraph, this graph has no 2-prime labeling (and
so the Petersen graph is not a line graph).

As expected, the concepts of k-prime integers and k-prime labelings can be looked
at in terms of subsets. For integers k and n with $2 \le k \le n$, a k-element subset of
$[n]$ is referred to as a *k-subset*. A *k-subset labeling* of a graph G is an assignment
of distinct k-subsets to the vertices of G such that the labels of two vertices u and
v have a nonempty intersection if and only if u and v are adjacent. The following
observation is immediate.

Observation 4.16 *For an integer $k \ge 2$, a graph G possesses a k-prime labeling if
and only if G has a k-subset labeling.*

As Figure 4.20 shows, the Petersen graph has a 3-subset labeling. While none of
the nine graphs of Figure 4.17 has a 2-subset labeling, each has a 3-subset labeling,
examples of which are shown in Figure 4.21 (where the set [12] of the first 12
positive integers is denoted by $\{1, 2, \ldots, 9, x, y, z\}$ and a 3-subset $\{a, b, c\}$ is once
again denoted by abc in Figure 4.21). In fact, every nontrivial connected graph has
a k-subset labeling for some integer $k \ge 2$, as we show next.

Theorem 4.17 *Every connected graph of order $n \ge 3$ has a k-subset labeling for
some integer k with $2 \le k < n$.*

Proof We proceed by induction on the order n of a connected graph. It is straight-
forward to show that the statement is true for all connected graphs of order 3 or 4.

Assume that every connected graph of order n, where $n \ge 4$, has a k-subset labeling for some integer k where $2 \le k < n$. Let G be a connected graph of order $n + 1$
and let v be a vertex of G that is not a cut-vertex of G, where $\deg_G v = d$. Thus, $G - v$
is a connected graph of order n. By the induction hypothesis, $G - v$ has a k-subset
labeling for some integer k with $2 \le k < n$. We may assume that all vertex labels
are k-subsets of $[r]$ for some positive integer r. Let $V(G) - \{v\} = \{v_1, v_2, \ldots, v_n\}$,

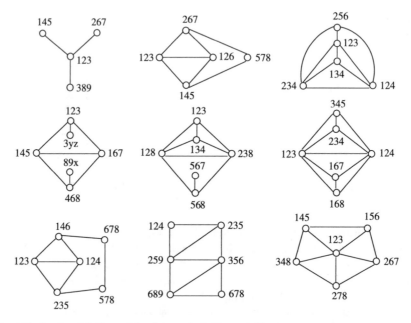

Fig. 4.21 3-subset labelings of the nine graphs in Figure 4.17

where $N_G(v) = \{v_1, v_2, \ldots, v_d\}$, and let S_i, $1 \le i \le n$, be the k-subset of $[r]$ assigned to v_i. Now, define

$$T_i = S_i \cup \{r + i\} \text{ for } 1 \le i \le n \text{ and } T = \{r + i : 1 \le i \le d\}.$$

Thus, $|T_i| = k + 1 \le n$ for $1 \le i \le n$ and $|T| = d \le n$.

* ⋆ If $d = k + 1$, then we assign T_i to v_i for $1 \le i \le n$ and T to v, producing a $(k + 1)$-subset labeling of G.
* ⋆ If $d < k + 1$, then let $T' = T \cup \{r + n + i : 1 \le i \le k + 1 - d\}$. Thus, $|T'| = d + (k + 1) - d = k + 1$. We assign T_i to v_i for $1 \le i \le n$ and T' to v, producing a $(k + 1)$-subset labeling of G.
* ⋆ If $d > k + 1$, then let $A = \{r + n + i : 1 \le i \le n(d - k - 1)\}$. Thus, $|A| = n(d - k - 1)$. Let $\{A_1, A_2, \ldots, A_n\}$ be a partition of A where $|A_i| = d - k - 1$ for $1 \le i \le n$. Define $T_i' = T_i \cup A_i$ for $1 \le i \le n$. Then $|T_i'| = (k + 1) + (d - k - 1) = d = |T|$ for $1 \le i \le n$. We assign T_i' to v_i for $1 \le i \le n$ and T to v, producing a d-subset labeling of G.   □

Theorem 4.17 suggests the following concept. For a connected graph G of order $n \ge 3$, let $\psi(G)$ denote the minimum positive integer k for which G has a k-subset labeling. Thus, if $\psi(G) = k$, then G has a j-subset labeling for every integer $j \ge k$ but no j-subset labeling for any integer j with $j < k$. Furthermore, if G is a graph with $\psi(G) = k \ge 3$ and v is a non-cut-vertex of G, then $\psi(G - v) \ge k - 1$.

Fig. 4.22 Two graphs G
with $\psi(G) = 4$ but
$\psi(G - v) = 3$ for every
non-cut-vertex v of G

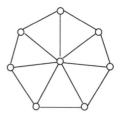

Necessarily, if G is a connected graph of order 3 or more with $\psi(G) = 2$ and v is a
non-cut-vertex of G, then $\psi(G - v) = 2$ as well. The graphs in Figure 4.21 all have
the property that $\psi(G) = 3$ and $\psi(G - v) = 2$ for every non-cut-vertex v of G.

The fact that none of the nine graphs of Figure 4.17 has a 2-subset labeling but
each has a 3-subset labeling suggests the following concept. For an integer $k \geq 2$, a
graph G is a *k-subset graph* if G possesses a k-subset labeling. Thus, a graph G is a
2-subset graph if and only if G is a line graph. So, k-subset graphs with $k \geq 3$ may
be considered generalizations of line graphs.

While the nine graphs of Figure 4.17 constitute a forbidden subgraph characterization of 2-subset graphs, this suggests the problem of determining the class $\mathscr{S}$ of
forbidden subgraphs for 3-subset graphs. Each such graph G in this class $\mathscr{S}$ then
has the property that $\psi(G) = 4$ but $\psi(G - v) = 3$ for every non-cut-vertex v of G.
Two members of $\mathscr{S}$ are shown in Figure 4.22.

It is useful to make one other comment in this chapter. Suppose that G is a k-subset
graph for some integer $k \geq 2$. Then G possesses a k-subset labeling f. Therefore, f
assigns distinct k-subsets to the vertices of G where two vertices are assigned non-
disjoint k-subsets if and only if they are adjacent. Consequently, for the complement
$\overline{G}$ of the graph G, the labeling f assigns disjoint k-subsets to two vertices of $\overline{G}$ if and
only if these vertices are adjacent in $\overline{G}$. Graphs with this property have been studied
by many where these labelings were referred to as *k-tuple colorings*. This coloring
concept was introduced by Saul Stahl [44] in 1976.

Chapter 5
Additive Labelings

In the preceding chapters, our interest has been on various types of vertex labelings of graphs, where typically the vertex labels are nonnegative integers. In this chapter, our emphasis initially shifts to edge labelings, where each edge labeling generates a vertex coloring possessing some desired property. Later in the chapter, we move on to vertex labelings that produce vertex colorings. All of these labelings have something in common—namely, the resulting vertex coloring is obtained by summing certain labels. In addition, rather than having distinct labels, as has often been the case, the goal here is to produce a desired coloring by permitting duplication of labels and minimizing the largest integer used as a label.

5.1 Rainbow Additive Labelings

It is well known (in fact part of graph theory folklore) that every nontrivial graph has at least two vertices having the same degree. Indeed, for every integer $n \geq 2$, there is only one connected graph G_n having exactly two vertices of the same degree. The graphs G_2, G_3, and G_4 are shown in Figure 5.1, where the vertices are labeled with their degrees. Of course, the complement $\overline{G}_n$ of G_n also has exactly two vertices of the same degree, although $\overline{G}_n$ is disconnected.

A graph G has been called *irregular* if no two vertices of G have the same degree. As we just stated, no nontrivial graph is irregular. A graph having exactly two vertices of the same degree is called *nearly irregular*. So, all three graphs of Figure 5.1 are nearly irregular. The situation is quite different with multigraphs, however. For example, each of the three multigraphs in Figure 5.2 is irregular. In each case, the vertices are labeled with their degrees. For $i = 3, 4$, the irregular multigraph M_i is obtained from the graph G_i of Figure 5.1 by the addition of a single edge.

The three multigraphs in Figure 5.2 can be represented in another manner. Figure 5.3 shows three graphs G_1, G_3, and G_4 (where, once again, G_3 and G_4 are the graphs in Figure 5.1). The edges of these three graphs are labeled with elements of the set [3]. In each case, the label assigned to an edge uv corresponds to the number

© The Author(s), under exclusive license to Springer Nature Switzerland AG 2019
G. Chartrand et al., *How to Label a Graph*, SpringerBriefs
in Mathematics, https://doi.org/10.1007/978-3-030-16863-6_5

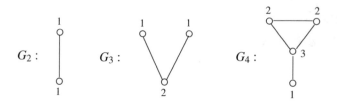

Fig. 5.1 Connected graphs having exactly two vertices of the same degree

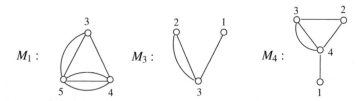

Fig. 5.2 Irregular multigraphs

of parallel edges joining u and v in the associated multigraph of Figure 5.2. From each such edge labeling, a vertex labeling is obtained by summing the labels of the edges incident with the vertex. Each vertex label therefore results in the degree of the vertex in the associated multigraph. This corresponds to an edge labeling which will be a major topic of discussion in this chapter.

An edge labeling $f : E(G) \to [k]$ of a graph G that results in the vertex labeling $f' : V(G) \to \mathbb{N}$ defined by

$$f'(v) = \sum_{u \in N(v)} f(uv)$$

is referred to as an *additive labeling* of G. If no two vertices of G have the same label, then f is a *rainbow additive labeling* of G. Thus, the labelings of the graphs in Figure 5.3 are rainbow additive labelings. The term *rainbow* is often used in graph theory to denote colorings where the vertices (or edges) in a graph or subgraph have different colors. That is, a rainbow additive labeling of a graph is an edge labeling

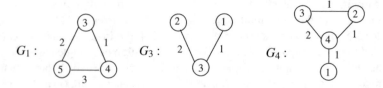

Fig. 5.3 The graphs G_1, G_2, G_3

that produces a rainbow vertex coloring. For this reason, we refer to a vertex coloring here rather than a vertex labeling.

A rainbow additive labeling of a graph G is similar to a type of labeling called an *antimagic labeling*. In the case of an antimagic labeling, however, if G has size m, then the edges are labeled with distinct elements of the set $[m]$.

If G is a connected graph of order 3 or more, then G always has a rainbow additive labeling. For example, if $E(G) = \{e_1, e_2, \ldots, e_m\}$, then the function $f : E(G) \rightarrow [2^{m-1}]$ defined by $f(e_i) = 2^{i-1}$ for $1 \leq i \leq m$ is one such rainbow additive labeling of G (since no two vertices have the same set of incident edges). The minimum positive integer k for which G has a rainbow additive labeling $f : E(G) \rightarrow [k]$ is the *rainbow additive index* ra(G) of G. This concept was introduced by Gary Chartrand in 1986 [6] at the 250th Anniversary of Graph Theory Conference held at Indiana University-Purdue University Fort Wayne (now called Purdue University Fort Wayne). At that time and in succeeding years, this concept has been studied and referred to using different terminologies. In fact, the rainbow additive index ra(G) of a graph G has often been called the *irregularity strength* of G, as the *strength* of a multigraph M is the maximum number of parallel edges joining any two vertices of M.

Since there is no irregular graph, ra$(G) \geq 2$ for every connected graph of order 3 or more. In fact, there is an infinite class of graphs G, already introduced, for which ra$(G) = 2$.

Proposition 5.1 *For every nearly irregular graph G_n of order $n \geq 3$, ra$(G_n) = 2$.*

Proof The rainbow additive labelings of the graphs G_3 and G_4 in Figure 5.3 show that ra$(G_3) = $ ra$(G_4) = 2$. Interchanging the labels 1 and 2 also results in a rainbow additive labeling of G_4 (see Figure 5.4). The labeling of G_5 in Figure 5.4 is also a rainbow additive labeling, which shows that ra$(G_5) = 2$.

For a nearly irregular graph G_n of order $n \geq 3$ with $V(G_n) = \{v_1, v_2, \ldots, v_n\}$ such that $\deg v_1 \leq \deg v_2 \leq \ldots \leq \deg v_n$, the degrees of these vertices are as follows:

$$\deg_{G_n} v_i = \begin{cases} i & \text{if } 1 \leq i \leq \lfloor \frac{n}{2} \rfloor \\ i - 1 & \text{if } \lfloor \frac{n}{2} \rfloor + 1 \leq i \leq n. \end{cases}$$

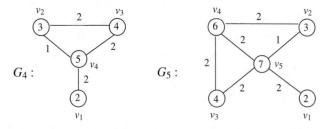

Fig. 5.4 Rainbow additive labelings of G_4 and G_5

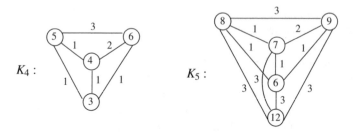

Fig. 5.5 Rainbow additive labelings of K_4 and K_5

In particular, the vertices $v_{\lfloor\frac{n}{2}\rfloor}$ and $v_{\lfloor\frac{n}{2}\rfloor+1}$ are the only two vertices of G_n having the same degree. The vertex v_n is adjacent to all other vertices of G_n. The labeling that assigns the label 2 to all edges except the edge $v_n v_{\lfloor\frac{n}{2}\rfloor}$, which is labeled 1, results in $f'(v_i) = 2\deg_{G_n} v_i$ for $i \neq \lfloor\frac{n}{2}\rfloor$, n and $f'(v_i) = 2\deg_{G_n} v_i - 1$ for $i = \lfloor\frac{n}{2}\rfloor$, n. Since the vertices v_i for $i \neq \lfloor\frac{n}{2}\rfloor$, n have distinct even degrees in G_n, they have distinct even colors. Since $f'(v_n) \neq f'(v_{\lfloor\frac{n}{2}\rfloor})$ and both colors are odd, this is a rainbow additive labeling of G_n and so $\mathrm{ra}(G_n) = 2$.   □

We saw that $\mathrm{ra}(K_3) = 3$. Actually, $\mathrm{ra}(K_n) = 3$ for every integer $n \geq 3$. Rainbow additive labelings showing that $\mathrm{ra}(K_4) \leq 3$ and $\mathrm{ra}(K_5) \leq 3$ are given in Figure 5.5. We will see in the next example that $\mathrm{ra}(K_4) = \mathrm{ra}(K_5) = 3$.

Proposition 5.2 *For every integer $n \geq 3$, $\mathrm{ra}(K_n) = 3$.*

Proof First, we show that $\mathrm{ra}(K_n) \geq 3$ for each integer $n \geq 3$. Suppose that there exists a rainbow additive labeling of K_n using only the labels 1 and 2 for some integer $n \geq 3$. Let G be the spanning subgraph of K_n whose edges are those labeled 1. Since no graph is irregular, there are two vertices u and v of G for which $\deg_G u = \deg_G v = k$ for some integer k. However then, the vertices u and v are both colored $1 \cdot k + 2(n - 1 - k) = 2n - k - 2$ in K_n, a contradiction.

It therefore remains to show that there is a rainbow additive labeling of every complete graph K_n, $n \geq 3$, with the integers 1, 2, 3. We have already seen this with K_3, K_4, and K_5. The labeling of K_4 is obtained from that of K_3 by adding a new vertex to K_3 and labeling each edge incident with this new vertex 1. The labeling of K_5 is obtained from that of K_4 by adding a new vertex to K_4 and labeling each edge incident with this new vertex 3. We continue this, alternating 1 and 3, as above. The vertex colors of K_5 are distinct integers in the range [6, 12]. Obtaining a labeling of K_6 in this manner, we have distinct vertex labels in the range [5, 13]. We can then proceed by induction to show that there is a rainbow additive labeling of K_n for all $n \geq 5$ such that the vertex colors are in the range $[n + 1, 3n - 5]$ if n is odd and are in the range $[n - 1, 3n - 3]$ if n is even.   □

By the same observation made in the proof of Proposition 5.2, it can be seen that $\mathrm{ra}(G) \geq 3$ for every regular graph G of order 3 or more. There is also an infinite class of regular graphs G for which $\text{ra}(G) = 4$. In particular, $\text{ra}(K_{r,r}) = 4$ for every odd integer $r \geq 3$. To illustrate this, we verify the following.

Proposition 5.3 $\text{ra}(K_{3,3}) = 4$.

Proof First, we show that there is no rainbow additive labeling of $G = K_{3,3}$ with the integers $1, 2, 3$. Assume, to the contrary, that there is such a labeling f. If we define a function f_1 on $E(G)$ by $f_1(e) = f(e) - 2$ for each edge e of G, then f_1 is a rainbow additive labeling of G with the integers $-1, 0, 1$. Let $U = \{u_1, u_2, u_3\}$ and $W = \{w_1, w_2, w_3\}$ be the partite sets of G. Since $\sum_{i=1}^{3} f_1'(u_i) = \sum_{i=1}^{3} f_1'(w_i)$, it follows that $\sum_{v \in V(G)} f_1'(v)$ is even. Since the set of vertex colors is a 6-element subset of $S = [-3, 3]$ and S contains four odd integers, the only integer that is not a vertex color of G is an even integer. In particular, $-3, -1, 1, 3$ are all vertex colors and -3 and 3 are colors of vertices in the same partite set, say $f_1'(u_1) = -3$ and $f_1'(u_2) = 3$. Therefore, every edge incident with u_1 is labeled -1 and every edge incident with u_2 is labeled 1. If there are two edges of u_3 having the same label, then two vertices of W have the same color, which cannot occur. Thus, the three edges incident with u_3 are labeled $-1, 0, 1$ and so $f'(u_3) = 0$. This, however, implies that $f'(w) = 0$ for some vertex $w \in W$, which is a contradiction. Since there is no rainbow additive labeling of G with $-1, 0, 1$, there is also none with $1, 2, 3$. As shown in Figure 5.6, there is a rainbow additive labeling of G with $1, 2, 3, 4$, however. Therefore, $\text{ra}(K_{3,3}) = 4$. $\square$

The rainbow additive indices of all regular complete bipartite graphs were determined in [6, 21].

Theorem 5.1 *For each integer $r \geq 2$,*

$$\text{ra}(K_{r,r}) = \begin{cases} 3 & if\ r\ is\ even \\ 4 & if\ r\ is\ odd. \end{cases}$$

With the single exception of the graphs $K_{r,r}$ where $r \geq 3$ is odd, all regular complete multipartite graphs have rainbow additive index 3 (see [13]).

Theorem 5.2 *For every regular complete k-partite graph G with $k \geq 3$,*

$$\text{ra}(G) = 3.$$

Fig. 5.6 A rainbow additive labeling of $K_{3,3}$

$K_{3,3}$:

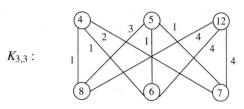

Fig. 5.7 A rainbow additive
labeling of $K_{3,4}$

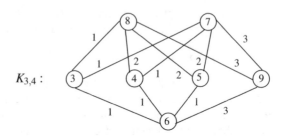

$K_{3,4}$:

While the rainbow additive index of $K_{3,3}$ is 4, the rainbow additive index of $K_{3,4}$ is smaller, namely, $\mathrm{ra}(K_{3,4}) = 3$. An appropriate labeling of $K_{3,4}$ is shown in Figure 5.7. In fact, $\mathrm{ra}(K_{3,s}) = 3$ for $s \in [4, 8]$.

While we have seen that $\mathrm{ra}(G)$ is either 3 or 4 for every regular complete multipartite graph G, there is no positive integer constant c such that $\mathrm{ra}(G) \le c$ for complete multipartite graphs G in general. In fact, for a fixed integer $r \ge 2$,

$$\lim_{s \to \infty} \mathrm{ra}(K_{r,s}) = \infty.$$

The value of the rainbow additive index of complete multipartite graphs that are not regular is an open problem.

Problem 5.1 Determine the value of $\mathrm{ra}(G)$ for complete multipartite graphs G that are not regular.

For all cycles and paths, the rainbow additive index has been determined.

Theorem 5.3 ([13]) *For an integer* $n \ge 3$,

$$\mathrm{ra}(C_n) = \begin{cases} \frac{n+1}{2} & \text{if } n \equiv 1 \pmod 4 \\ \frac{n+2}{2} & \text{if } n \text{ is even} \\ \frac{n+3}{2} & \text{if } n \equiv 3 \pmod 4. \end{cases} \tag{5.1}$$

Theorem 5.4 ([6]) *For an integer* $n \ge 3$,

$$\mathrm{ra}(P_n) = \begin{cases} \frac{n}{2} & \text{if } n \equiv 0 \pmod 4 \\ \frac{n+1}{2} & \text{if } n \text{ is odd} \\ \frac{n+2}{2} & \text{if } n \equiv 2 \pmod 4. \end{cases}$$

It is immediate that the rainbow additive index of a star of order $n \ge 3$ is $n - 1$. Not only do paths and stars of large order have large rainbow additive index, but this is true for all trees of large order.

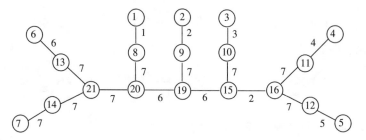

Fig. 5.8   A tree T of order 19 with $ra(T) = \frac{19+2}{3} = 7$

Theorem 5.5 ([1, 6]) *If T is a tree of order $n \geq 3$ that is not a star, then*

$$\frac{n+2}{3} \leq ra(T) \leq n - 2.$$

The upper bound in Theorem 5.5 was established in [1], while the lower bound was determined in [6]. Both bounds are sharp. If T is a double star (a tree of diameter 3) of order $n \geq 4$, then $ra(T) = n - 2$. The tree T of Figure 5.8 shows that the lower bound is sharp when $n = 19$.

5.2   Proper Additive Labelings

As we saw in the preceding section, a rainbow additive labeling of a graph G is an edge labeling of G that gives rise to a rainbow vertex coloring of G. There are also edge labelings of graphs that give rise to other types of vertex colorings. Indeed, rainbow vertex colorings are not even the most popular type of vertex coloring. Without a doubt, the best known vertex colorings are the proper colorings (where every two adjacent vertices are required to have different colors). Interest in these vertex colorings originated from attempts to solve the famous Four Color Problem. An additive labeling that results in a proper vertex coloring is a *proper additive labeling*. Since every connected graph of order 3 or more has a rainbow additive labeling, it also has a proper additive labeling. The minimum positive integer k for which a graph G has a proper additive labeling using elements from the set $[k]$ is called the *proper additive index* of G, denoted by $pa(G)$. Clearly, $pa(G) \leq ra(G)$ for every connected graph G of order 3 or more.

The proper additive index of a graph G is therefore the minimum strength of a multigraph M obtained from G by possibly adding additional edges between adjacent vertices of G so that every two adjacent vertices of M have different degrees. If every two adjacent vertices of G already have different degrees, then $pa(G) = 1$. For example, $pa(K_{r,s}) = 1$ when $r \neq s$. See Figure 5.9 for the graph $H_1 = K_{2,3}$. If this is not the case, then $pa(G) \geq 2$. The graph $H_2 = C_4$, also shown in Figure 5.9,

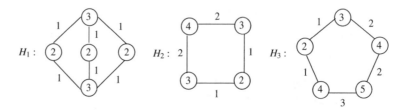

Fig. 5.9 Graphs with proper additive indices 1, 2, and 3

has proper additive index 2. If one were to label the edges of $H_3 = C_5$ only with the labels 1 and 2, then there is at least one edge of H_3 whose two neighboring edges have the same label, resulting in two adjacent vertices of the same color, which implies that $pa(H_3) \geq 3$. That $pa(H_3) = 3$ is shown in Figure 5.9 as well.

The interesting feature of proper additive labelings is that no example of a connected graph of order 3 or more has been found for which it is necessary to use the edge label 4. A popular conjecture (due to Michal Karoński, Tomasz Łuczak, and Andrew Thomason) resulting from this fact is the following, which goes by a rather catchy name.

The 1-2-3 **Conjecture** *If G is a connected graph of order 3 or more, then there is a proper additive labeling of G using the labels 1, 2, 3.*

Karoński, Łuczak, and Thomason [26] showed that this conjecture holds for all connected graphs of order 3 or more having chromatic number at most 3.

Theorem 5.6 *If G is a connected graph of order 3 or more having chromatic number 2 or 3, then $pa(G) \leq 3$.*

While it has not been shown that $pa(G) \leq 3$ for every connected graph G of order 3 or more, it has not even been shown that $pa(G) \leq 4$ for such graphs. The following result, due to Maciej Kalkowski, Michal Karoński, and Florian Pfender [25], has been established, however.

Theorem 5.7 *If G is a connected graph of order 3 or more, then $pa(G) \leq 5$.*

5.3   Monochromatic Additive Labelings

In addition to rainbow colorings and proper colorings, the third popular coloring in graph theory is a *monochromatic coloring*, which often occurs with edge colorings in Ramsey Theory. A *monochromatic additive labeling* of a graph G is an additive labeling that results in a vertex coloring in which all vertices of G have the same color. A monochromatic additive labeling of a graph G is therefore equivalent to adding additional edges between adjacent vertices of G, if necessary, to produce a

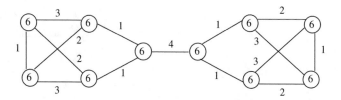

Fig. 5.10 A regular graph with a monochromatic additive labeling

regular multigraph. If G itself is regular, any labeling that assigns the same positive
integer to every edge of G is a monochromatic additive labeling.

A 1-*factor* of a graph G is a spanning 1-regular subgraph of G. A graph G is
said to be 1-*factorable* if there are pairwise edge-disjoint 1-factors $F_1, F_2, \ldots, F_r$
of G such that $E(G) = \cup_{i=1}^{r} E(F_i)$. Indeed, if G is a 1-factorable r-regular graph
with 1-factors F_i $(1 \leq i \leq r)$, then the labeling that assigns each edge of F_i the label
$a_i \in \mathbb{N}$ is a monochromatic additive labeling in which every vertex color is $\sum_{i=1}^{r} a_i$.
A regular graph that is not 1-factorable may very well have a monochromatic additive
labeling in which not all edges are labeled the same (see Figure 5.10).

A major difference between monochromatic additive labelings and rainbow or
proper additive labelings is that not all non-regular graphs have a monochromatic
additive labeling. For example, no tree of order 3 or more has such a labeling and $K_{r,s}$
with $r \neq s$ also has no such labeling. If G is a graph possessing a monochromatic
additive labeling, then the minimum positive integer k for which G has a monochromatic additive labeling using elements from the set $[k]$ is called the *monochromatic
additive index* of G, denoted by $\text{ma}(G)$. Therefore, $\text{ma}(G) = 1$ if and only if G is
regular.

While no non-regular complete bipartite graph has a monochromatic additive
labeling, a non-regular complete k-partite graph with $k \geq 3$ may have a monochromatic additive labeling. For example, the graph $K_{2,3,4}$ has a monochromatic additive
labeling. To see this, let V_1, V_2, V_3 be the partite sets of $K_{2,3,4}$ where $|V_1| = 2$,
$|V_2| = 3$, and $|V_3| = 4$. For $i, j \in \{1, 2, 3\}$ and $i \neq j$, let $[V_i, V_j]$ denote the set of
edges joining a vertex of V_i and a vertex of V_j. Define an edge labeling f of $K_{2,3,4}$
by assigning the label 4 to each edge in $[V_1, V_2]$, the label 9 to each edge in $[V_1, V_3]$,
and the label 10 to each edge in $[V_2, V_3]$ (see Figure 5.11). Since the color of every
vertex of $K_{2,3,4}$ is 48, it follows that f is a monochromatic additive labeling of $K_{2,3,4}$.
As another example, the graph $K_{3,4,5}$ also has a monochromatic additive labeling.
Let V_1, V_2, V_3 be the partite sets of $K_{3,4,5}$ with $|V_1| = 3$, $|V_2| = 4$, and $|V_3| = 5$.
Define an edge labeling f of $K_{3,4,5}$ by assigning the label 5 to each edge in $[V_1, V_2]$,
the label 8 to each edge in $[V_1, V_3]$, and the label 9 to each edge in $[V_2, V_3]$ (see
Figure 5.11). Since the color of every vertex of $K_{3,4,5}$ is 60, it follows that f is a
monochromatic additive labeling of $K_{3,4,5}$. Whether there is a monochromatic additive labeling of these two graphs where the labels of the edges joining vertices in each
pair of partite sets are not a constant is not known. In any case, the monochromatic
additive labelings in Figure 5.11 show that $\text{ma}(K_{2,3,4}) \leq 10$ and $\text{ma}(K_{3,4,5}) \leq 9$.

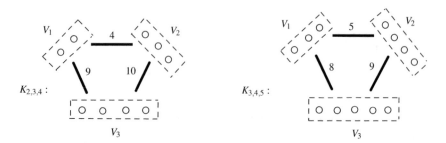

Fig. 5.11 Monochromatic additive labelings of $K_{2,3,4}$ and $K_{3,4,5}$

The graphs $K_{2,3,4}$ and $K_{3,4,5}$ belong to a class of complete 3-partite graphs described in the following example.

Proposition 5.4 *The complete 3-partite graph $K_{r,s,t}$ with $2 \le r \le s \le t$ has a monochromatic additive labeling if and only if $t < r + s$.*

Proof Let V_1, V_2, V_3 be the partite sets of $G = K_{r,s,t}$ where $|V_1| = r$, $|V_2| = s$, and $|V_3| = t$. Suppose that G has a monochromatic additive labeling f. Then f produces a monochromatic vertex coloring of G in which each vertex color is some positive integer p. The set of edges incident with the vertices of V_3 is $[V_1 \cup V_2, V_3]$ and the set of edges incident with the vertices of $V_1 \cup V_2$ is $[V_1, V_2] \cup [V_1 \cup V_2, V_3]$. Since the sum of the labels of the edges incident with the vertices of V_3 is pt and the sum of the labels of the edges in $[V_1 \cup V_2, V_3]$ incident with the vertices of $V_1 \cup V_2$ is less than $rp + sp$, it follows that $tp < (r + s)p$ and so $t < r + s$.

It remains to show that if $t < r + s$, then G has a monochromatic additive labeling. Since $t < r + s$, it follows that $r + s - t$ is a positive integer. Define an edge labeling f of G by assigning the label $a = t(r + s - t)$ to each edge in $[V_1, V_2]$, the label $b = s(r + t - s)$ to each edge in $[V_1, V_3]$, and the label $c = r(s + t - r)$ to each edge in $[V_2, V_3]$. Since the resulting color of every vertex of G is $2rst$, it follows that f is a monochromatic additive labeling of G. If $d = \gcd(a, b, c)$, then multiplying each edge label by $1/d$ results in another monochromatic additive labeling of G. This implies that $\mathrm{ma}(G) \le r(s - r + t)/d$.   □

The complete 3-partite graphs considered in Proposition 5.4 have another well-known property. A path in a graph G is a *Hamiltonian path* if it contains every vertex of G. A graph G is *Hamiltonian-connected* if every two vertices of G are connected by a Hamiltonian path. A *Hamiltonian cycle* in a graph G is a cycle containing every vertex of G. A graph having a Hamiltonian cycle is a *Hamiltonian graph*. For a complete k-partite graph $G = K_{n_1,n_2,\ldots,n_k}$, where $k \ge 3$ and $n_1 \le n_2 \le \cdots \le n_k$, it is well known that G is Hamiltonian if and only if $n_k \le \sum_{i=1}^{k-1} n_i$ and G is Hamiltonian-connected if and only if $n_k < \sum_{i=1}^{k-1} n_i$. Not all non-regular complete multipartite graphs have a monochromatic additive labeling. By an argument similar to the one in the proof of Proposition 5.4, we can show that if $G = K_{n_1,n_2,\ldots,n_k}$ is a complete k-partite graph, where $k \geq 3, n_1 \leq n_2 \leq \cdots \leq n_k$ and $n_k \geq \sum_{i=1}^{k-1} n_i$, then G has no monochromatic additive labeling. Thus, we have the following conjecture.

Conjecture 5.1 A complete k-partite graph, $k \geq 3$, has a monochromatic additive labeling if and only if it is Hamiltonian-connected.

This brings up the following question.

Problem 5.2 If G is a complete multipartite graph having a monochromatic additive labeling, then what is $ma(G)$?

On the other hand, the more general problem is the following.

Problem 5.3 Which graphs have monochromatic additive labelings?

5.4 Proper Sigma Labelings

We now consider some additional classes of additive labelings which, in this case, deal with vertex labelings rather than edge labelings. Among the variety of ways in which the famous Four Color Theorem can be stated, perhaps the best known (in terms of graphs) is the following.

The chromatic number of every planar graph is at most 4.

That is, no more than four colors are needed if distinct colors are to be assigned to adjacent vertices of a planar graph. In 2008, Gary Chartrand introduced (both in the book [9] and at the Twenty-Second Midwest Conference on Combinatorics, Cryptography, and Computing) four new vertex colorings (labelings) of graphs, where adjacent vertices may be colored the same and yet adjacent vertices are distinguished in some way by the coloring. In each case, no more than four colors are needed to distinguish adjacent vertices in any planar graph. One of these colorings is the vertex version of additive labelings.

Let G be a nontrivial connected graph and let $f : V(G) \to \mathbb{N}$ be a vertex labeling of G. The labeling f gives rise to the vertex coloring $\sigma : V(G) \to \mathbb{N}$ defined by

$$\sigma(v) = \sum_{u \in N(v)} f(u). \tag{5.2}$$

If $\sigma(x) \neq \sigma(y)$ whenever $xy \in E(G)$, then f is called a *proper sigma labeling* or simply a *sigma labeling* of G, as used in [7]. The minimum positive integer k for which $f(v) \in [k]$ for all $v \in V(G)$ in some sigma labeling f of G is called the *sigma chromatic number* of G, denoted by $\sigma(G)$. Therefore, $\sigma(G) = 1$ if and only if every

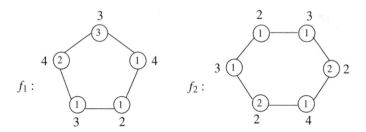

Fig. 5.12 Sigma labelings of C_5 and C_6

two adjacent vertices of G have different degrees. For example, $\sigma(K_{r,s}) = 1$ for all positive integers r and s with $r \neq s$. Later, sigma labelings have been studied using different terminologies, including calling this labeling a *lucky labeling*.

Figure 5.12 shows a sigma labeling f_1 of the 5-cycle C_5 and a sigma labeling f_2 of the 6-cycle C_6, where $f_i(v)$ ($i = 1, 2$) is placed inside a vertex v and $\sigma(v)$ is placed outside v. The sigma labeling f_1 of C_5 uses the labels 1, 2, 3 and the sigma labeling f_2 of C_6 uses the labels 1 and 2. Since C_6 is regular, it follows that $\sigma(C_6) \geq 2$. The sigma labeling f_2 of C_6 shows that $\sigma(C_6) \leq 2$ and so $\sigma(C_6) = 2$. For the cycle C_5, it turns out that $\sigma(C_5) = 3$. Since C_5 is regular, $\sigma(C_5) \geq 2$. The sigma labeling f_1 in Figure 5.12 shows that $\sigma(C_5) \leq 3$. Assume, to the contrary, that $\sigma(C_5) = 2$. Let $C_5 = (v_1, v_2, \ldots, v_5, v_1)$ and let f be a sigma labeling of C_5 using the labels 1 and 2. Then two adjacent vertices of C_5 must be labeled the same, say $f(v_1) = f(v_2) = 1$. Since $\sigma(v_1) \neq \sigma(v_2)$, it follows that $\{f(v_3), f(v_5)\} = \{1, 2\}$, say $f(v_3) = 1$ and $f(v_5) = 2$. If $f(v_4) = 1$, then $\sigma(v_2) = \sigma(v_3) = 2$; while if $f(v_4) = 2$, then $\sigma(v_4) = \sigma(v_5) = 3$, which is impossible. Therefore, $\sigma(C_5) = 3$.

One of the key features of the sigma chromatic number of a graph is that it never exceeds its chromatic number (see [7]). Therefore, by the Four Color Theorem, $\sigma(G) \leq 4$ for every planar graph G.

Theorem 5.8 *For every graph G, $\sigma(G) \leq \chi(G)$.*

It is possible that $\sigma(G) = \chi(G)$ for a graph G. For example, $\sigma(C_n) = \chi(C_n)$ for every integer $n \geq 3$. That is, $\sigma(C_n) = 2$ if n is even and $\sigma(C_n) = 3$ if n is odd. Furthermore, every pair a, b of positive integers with $a \leq b$ can be realized as the sigma chromatic number and chromatic number, respectively, of some connected graph (see [7]).

Theorem 5.9 *For every pair a, b of positive integers with $a \leq b$, there exists a connected graph G such that $\sigma(G) = a$ and $\chi(G) = b$.*

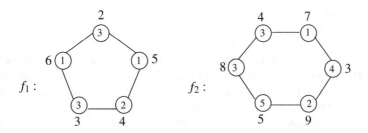

Fig. 5.13 Rainbow sigma labelings of C_5 and C_6

5.5 Rainbow Sigma Labelings

Let G be a nontrivial connected graph and let $f : V(G) \to \mathbb{N}$ be a vertex labeling of G. If f gives rise to the vertex coloring $\sigma : V(G) \to \mathbb{N}$ defined by (5.2) such that $\sigma(x) \neq \sigma(y)$ for every two distinct vertices x and y of G, then f is called a *rainbow sigma labeling* of G. The minimum positive integer k for which $f(v) \in [k]$ for all $v \in V(G)$ in some rainbow sigma labeling f of G is called the *rainbow sigma chromatic number* of G, denoted by $\sigma_r(G)$. Thus, $\sigma(G) \leq \sigma_r(G)$ for every nontrivial connected graph G.

Figure 5.13 shows a rainbow sigma labeling f_1 of C_5 and a rainbow sigma labeling f_2 of C_6. Since $\sigma(C_5) = 3$, the rainbow sigma labeling f_1 of C_5 shows that $\sigma_r(C_5) = 3$. The rainbow sigma labeling f_2 of C_6 in Figure 5.13 shows that $\sigma_r(C_6) \leq 5$.

In fact, $\sigma_r(C_6) = 5$. Since C_6 is a regular graph of order 6, in any rainbow sigma labeling of C_6, there is a vertex v such that $\sigma(v) \geq 7$ and so $\sigma_r(C_6) \geq 4$. Assume, to the contrary, that $\sigma_r(C_6) = 4$. Let there be given a rainbow sigma labeling f of C_6 using the labels in [4]. Since no two vertices of v_1, v_3, v_5 can be labeled the same and no two vertices of v_2, v_4, v_6 can be labeled the same, it follows that $|\{\sigma(v_1), \sigma(v_3), \sigma(v_5)\}| = |\{\sigma(v_2), \sigma(v_4), \sigma(v_6)\}| = 3$. Therefore, $|\{\sigma(v_1), \sigma(v_3), \sigma(v_5)\} \cap \{\sigma(v_2), \sigma(v_4), \sigma(v_6)\}| \geq 2$. Hence, there are $a, b \in [4]$ with $a \neq b$ such that $a, b \in \{\sigma(v_1), \sigma(v_3), \sigma(v_5)\} \cap \{\sigma(v_2), \sigma(v_4), \sigma(v_6)\}$. However then, there exist distinct vertices v_i and v_j of C_6 with $\sigma(v_i) = \sigma(v_j) = a + b$, a contradiction. Therefore, $\sigma_r(C_6) = 5$.

Among the many problems in this area of research is the following.

Problem 5.4 Determine the rainbow sigma chromatic number of cycles.

Unlike (proper) sigma labelings, rainbow sigma labelings may not exist for some graphs. Two vertices u and v in a connected graph G are called *twins* if u and v have the same neighborhood (set of neighbors). If G contains twins u and v, then $N(u) = N(v)$ in G and so $\sigma(u) = \sigma(v)$ for any additive vertex labeling of G. Thus, there is no rainbow sigma labeling of G. For example, if r and s are positive integers with $r \neq s$, then $K_{r,s}$ contains twins and so there is no rainbow sigma labeling of $K_{r,s}$. For every nontrivial connected graph without twins, such labelings always exist.

Proposition 5.5 *A nontrivial connected graph G has a rainbow sigma labeling if and only if G has no twins.*

Proof We have seen that if G contains twins, then G has no rainbow sigma labeling. Next, let G be a connected graph of order $n \geq 2$ that contains no twins. Let $V(G) = \{v_1, v_2, \ldots, v_n\}$. Define the labeling $f : V(G) \rightarrow \mathbb{N}$ by $f(v_i) = 2^{i-1}$ for $1 \leq i \leq n$. Since G has no twins, no two vertices of G have the same neighbors. It follows that $\sigma(u) \neq \sigma(v)$ for every two distinct vertices u and v and so f is a rainbow sigma coloring of G.   □

5.6 Monochromatic Sigma Labelings

Let G be a nontrivial connected graph and let $f : V(G) \rightarrow \mathbb{N}$ be a vertex labeling of G. If f gives rise to the vertex coloring $\sigma : V(G) \rightarrow \mathbb{N}$ defined by (5.2) such that $\sigma(x) = \sigma(y)$ for every two vertices x and y of G, then f is called a *monochromatic sigma labeling*. The minimum positive integer k for which $f(v) \in [k]$ for all $v \in V(G)$ in some sigma labeling f of G is called the *monochromatic sigma chromatic number* of G, denoted by $\sigma_m(G)$. For example, let $G = K_{2,3,4}$ with partite sets V_1, V_2, V_3 where $|V_1| = 2$, $|V_2| = 3$, and $|V_3| = 4$. Define the vertex labeling f by assigning the label 6 to each vertex in V_1, the label 4 to each vertex in V_2, and the label 3 to each vertex in V_3 (see Figure 5.14). Since $\sigma(v) = 24$ for each vertex v of G, it follows that f is a monochromatic sigma labeling and $\sigma_m(G) \leq 6$. This example illustrates the following fact.

Proposition 5.6 *For integers r, s, t with $2 \leq r \leq s \leq t$, the complete 3-partite graph $K_{r,s,t}$ has a monochromatic sigma labeling.*

Proof Let $G = K_{r,s,t}$ with partite sets V_1, V_2, V_3 where $|V_1| = r, |V_2| = s, |V_3| = t$, and $d = \gcd(rs, rt, st)$. Define the vertex labeling f by assigning the label st/d to each vertex in V_1, the label rt/d to each vertex in V_2, and the label rs/d to each vertex in V_3. Since $\sigma(v) = 2rst/d$ for each vertex v of G, it follows that f is a monochromatic sigma labeling and $\sigma_m(G) \leq st/d$.   □

Fig. 5.14 A monochromatic sigma labeling of $K_{2,3,4}$

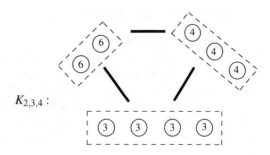

$K_{2,3,4}$:

Proposition 5.6 can be generalized to obtain the following.

Proposition 5.7 *Let $G = K_{n_1,n_2,\ldots,n_k}$ be a complete k-partite graph with $n_1 \leq n_2 \leq \cdots \leq n_k$ and $k \geq 3$ and let $\gamma = \prod_{i=1}^{k} n_i$ and $d = \gcd\left(\frac{\gamma}{n_1}, \frac{\gamma}{n_2}, \ldots, \frac{\gamma}{n_k}\right)$. Then*

$$\sigma_m(G) \leq \frac{(k-1)}{d} \prod_{i=2}^{k} n_i.$$

Among the problems dealing with this concept is the following.

Problem 5.5 Determine $\sigma_m(G)$ for graphs G belonging to well-known classes of graphs.

Chapter 6
Zonal Labelings

A vertex labeling of planar graphs using the nonzero elements of $\mathbb{Z}_3$ is introduced having the property that the sum of the labels of the vertices on the boundary of each zone (or region) is zero in $\mathbb{Z}_3$. We also describe a connection between this labeling and the famous Four Color Theorem.

6.1 Zonal Graphs

In each of the additive labelings described in Chapter 5, our goal was to minimize the largest label in a labeling that produces a particular vertex coloring. In this chapter, our goal is to describe another vertex labeling where very few labels are available. First, we consider the trees of order 6 we encountered earlier (see Figure 6.1) and assign to each vertex a single label, namely, the label 1.

If we were to add the labels of all six vertices in each tree of Figure 6.1, we obtain the sum 6, which, of course, is not surprising since all six trees have order 6. If we were to think of the label 1 as an element of $\mathbb{Z}_3$, then the sum in $\mathbb{Z}_3$ of the labels in each tree of order 6 is 0. However, this is not the case for trees of order 7 or 8 for example. On the other hand, if we were to allow ourselves the option of assigning every vertex of a nontrivial tree either of the two labels 1 and 2 of $\mathbb{Z}_3$, then, here too, we can always obtain a sum of 0 in $\mathbb{Z}_3$.

Proposition 6.1 *If T is a nontrivial tree, then there exists a labeling of the vertices of T with the elements 1 and 2 of $\mathbb{Z}_3$ so that the sum (in $\mathbb{Z}_3$) of the labels of all vertices of T is 0.*

Proof Since T is a nontrivial tree, T has order n for some integer $n \geq 2$. If n is even, then we can assign the label 1 to half of the vertices of T and the label 2 to the other half, giving us a sum of 0 in $\mathbb{Z}_3$. Suppose, then, that $n \geq 3$ is odd. Thus, $n = 2k + 1$ for some positive integer k. Then $n = 2k + 1 = (k - 1) + (k + 2)$. If we assign the label 1 to $k + 2$ vertices of T and assign the label 2 to the other $k - 1$ vertices of T

© The Author(s), under exclusive license to Springer Nature Switzerland AG 2019
G. Chartrand et al., *How to Label a Graph*, SpringerBriefs
in Mathematics, https://doi.org/10.1007/978-3-030-16863-6_6

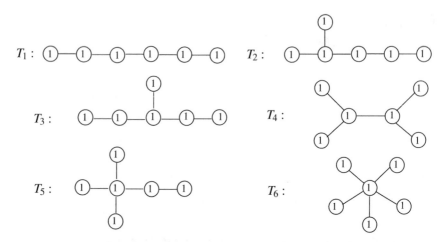

Fig. 6.1 The six trees of order 6

Fig. 6.2 The graph K_3

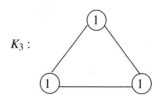

and add these labels, we have $1 \cdot (k + 2) + 2(k - 1) = (k + 2) + 2k - 2 = 3k = 0$ in $\mathbb{Z}_3$.   □

So, what happens here is that all three elements of $\mathbb{Z}_3$ are being used as labels, namely, 0, 1, and 2. In particular, the vertices of a tree are labeled either 1 or 2 and something else is labeled 0. But what exactly is labeled 0?

Next, let's look at a graph that is not a tree, say K_3. If we label each vertex of K_3 with the label 1, we obtain a sum of 3 (or 0 in $\mathbb{Z}_3$). See Figure 6.2.

Let's now look at n-cycles for an arbitrary integer $n \geq 3$. Of course, every tree and every cycle is a planar graph. When a tree is embedded in the plane, there is only one zone (region). However, every n-cycle C_n (embedded in the plane) divides the plane into two zones, where the boundary of each zone is the same, namely, C_n itself. Each vertex of C_n can be labeled 1 or 2 (in $\mathbb{Z}_3$) so that the sum (in $\mathbb{Z}_3$) of the labels of the vertices on the boundary of every zone is 0. See Figure 6.3 for C_3, C_4, and C_5.

This suggests something more general. Let G be a connected planar graph embedded in the plane (resulting in a plane graph) and suppose that each vertex of G is labeled 1 or 2 in $\mathbb{Z}_3$. Let R be a zone of G. Define the *label of R* as the sum in $\mathbb{Z}_3$ of the labels of the vertices on the boundary of R.

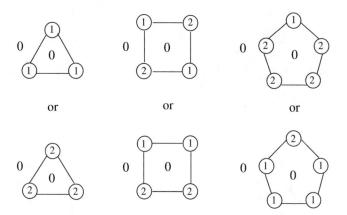

Fig. 6.3 Some n-cycles C_n

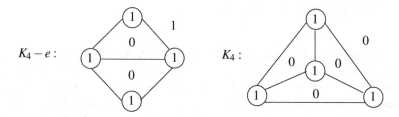

Fig. 6.4 The planar graphs $K_4 - e$ and K_4

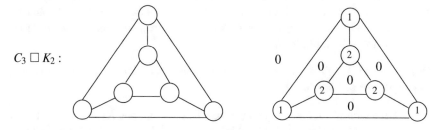

Fig. 6.5 The graph $C_3 \square K_2$

The planar graphs $K_4 - e$ and K_4 are more interesting as they have three and four zones, respectively, when these graphs are embedded in the plane. We can obtain a label of 0 for all zones in the case of K_4; but for $K_4 - e$, we can obtain a label of 0 for the interior zones and the label 1 (or 2) for the exterior zone, drawn as indicated. See Figure 6.4. In fact, there is no way to obtain the label 0 for all three zones of $K_4 - e$.

The planar graph $C_3 \square K_2$ has five zones. We can label each vertex of this graph either 1 or 2 so that the label of every zone is 0. See Figure 6.5.

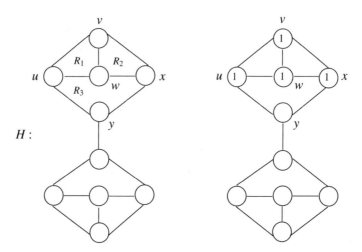

Fig. 6.6 A cubic plane graph that has no zonal labeling

All of this gives rise to a special kind of labeling. A connected plane graph G (or multigraph) is said to have a *zonal labeling* if the vertices of G can be labeled with the elements 1 and 2 of $\mathbb{Z}_3$ in such a way that the label of each zone is 0. If G admits a zonal labeling, then G is *zonal*. These concepts were introduced in 2014 by Cooroo Egan. We have seen that every tree, every cycle, and the graphs K_4 and $C_3 \,\Box\, K_2$ are zonal, but $K_4 - e$ is not.

An observation will be useful to us here. Let ℓ be a labeling of the vertices of a graph G with the labels 1 and 2 of $\mathbb{Z}_3$. The vertex labeling $\overline{\ell}$ of G defined by $\overline{\ell}(v) = 3 - \ell(v)$ for each vertex v of G is called the *complementary labeling* of G. The following is then immediate.

Observation 6.1 *If ℓ is a zonal labeling of a connected plane graph, then so too is its complementary labeling $\overline{\ell}$.*

We have seen in Figures 6.4 and 6.5 that the cubic graphs K_4 and $C_3 \,\Box\, K_2$ are both zonal. We now show that the cubic graph $C_n \,\Box\, K_2$ is zonal for every integer $n \geq 3$. Let $C = (u_1, u_2, \ldots, u_n, u_1)$ and $C' = (v_1, v_2, \ldots, v_n, v_1)$ be two n-cycles. Then $C_n \,\Box\, K_2$ can be constructed from the two cycles C and C' by adding the edges $u_i v_i$ for $1 \leq i \leq n$. Let ℓ be a zonal labeling of C. This zonal labeling of C can be extended to a zonal labeling ℓ of $C_n \,\Box\, K_2$ by defining $\ell(v_i) = \overline{\ell}(u_i)$ for $1 \leq i \leq n$.

Let's look at another cubic planar graph, namely, the graph H shown in Figure 6.6. Suppose that there exists a zonal labeling of H. Since the boundary of the zone R_1 is a 3-cycle C, the vertices u, v, w of C must all be labeled 1 (or all labeled 2), say 1. Thus, x must be labeled 1 as well. Since the boundary of the zone R_3 is a 4-cycle, the only way for R_3 to be labeled 0 is if two of the vertices u, w, x, y are labeled 1 and the other two vertices are labeled 2. However, three of these vertices are already labeled 1. So, there is no zonal labeling of H. That is, H is not zonal. Of course, the

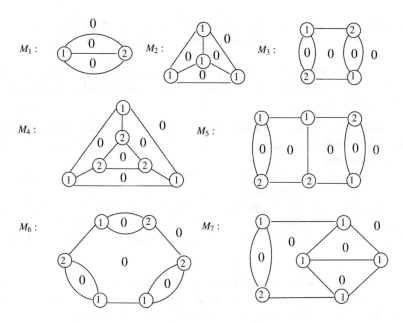

Fig. 6.7 Cubic bridgeless plane graphs and multigraphs of order at most 6

cubic graph H contains a bridge. This suggests the following question.

Is it the bridge in H that causes H not to have a zonal labeling?

Let's look at all seven bridgeless cubic planar graphs (or multigraphs) of order at most 6, to determine which of these are zonal. See Figure 6.7. This results in the following observation.

Observation 6.2 *All bridgeless cubic plane graphs and multigraphs of order at most 6 are zonal.*

These and other examples seem to suggest the following conjecture, which we will visit again later in the chapter.

Conjecture 6.1 A connected cubic plane graph (or multigraph) G is zonal if and only if G is bridgeless.

Among the zonal graphs and multigraphs we have seen thus far, it therefore appears to be of interest to consider connected bridgeless cubic plane graphs (or multigraphs) for the present, which we refer to as *cubic maps*. This class of graphs may be familiar as it has occurred often in graph theory literature in other contexts.

Francis Guthrie is credited with introducing (in 1852) perhaps the most famous problem in graph theory and one of the most famous in all of mathematics.

The Four Color Problem Can the regions (zones) of every plane graph be colored with four or fewer colors so that every two regions with a common boundary line are colored differently?

This problem was eventually solved 124 years later, in 1976, which resulted in the following theorem, a computer-aided proof of which was given by Kenneth Appel and Wolfgang Haken.

The Four Color Theorem The regions of every plane graph can be colored with four or fewer colors so that every two regions with a common boundary line are colored differently.

Many mathematicians (and non-mathematicians) tried to solve this problem prior to 1976 using a variety of methods. Among the many unsuccessful attempts to solve the Four Color Problem was one that occurred in 1880, due to the Scottish mathematician Peter Tait. Even at that time it was known that the Four Color Problem could be solved if it could be solved for all cubic maps. Tait proved the following theorem.

Theorem 6.3 (Tait's Theorem) *The regions of a cubic map M can be colored with four or fewer colors so that every two adjacent regions are colored differently if and only if the edges of M can be colored with three colors so that every two adjacent edges are colored differently, producing a proper edge coloring with three colors (that is, the chromatic index of M is $\chi'(M) = 3$).*

Tait thought that his theorem would lead to a proof of the Four Color Theorem as he thought it would be easy to prove that there is a proper coloring of the edges of every bridgeless cubic planar graph with three colors. But, as it turned out, this problem is equivalent to, and therefore just as difficult as, the Four Color Problem. Of course, since it is now known that the Four Color Theorem is true, it is also known that the edges of every cubic map M can, in fact, be colored with three colors such that every two adjacent edges of M are colored differently.

6.2 Zonal Labelings of Cycles

As it turns out, there is, in fact, a connection between proper 3-colorings of the edges of cubic maps and zonal labelings of these graphs. In order to describe this connection, we first consider proper edge colorings of cycles using colors from the set $\{1, 2, 3\}$. If the order n of a cycle is odd, then all three of these colors must be used; while if n is even, then all three of these colors may be used. We represent an n-cycle C, $n \geq 3$, as follows:

$$C = (v_1, v_2, \ldots, v_{n-1}, v_n, v_{n+1} = v_1),$$

where we assume that C is drawn in the plane as indicated in Figure 6.8.

In a proper edge coloring c of the n-cycle C, the color of an edge $v_i v_{i+1}$ $(1 \leq i \leq n)$ of C is then denoted by $c(v_i v_{i+1})$, which, therefore, is one of the colors $1, 2$, or 3. For a proper coloring c of the edges of a cycle with colors from the set $\{1, 2, 3\}$, the coloring $\bar{c}$ of the edges of C defined by $\bar{c}(e) = 4 - c(e)$ for each edge e of C is called the *complementary coloring* of c. An important characteristic of the complementary coloring is the following.

Observation 6.3 *The complementary coloring of a proper 3-coloring of the edges of a cycle C with colors from the set $\{1, 2, 3\}$ is also a proper 3-coloring of C with colors from $\{1, 2, 3\}$.*

For such a cycle C drawn in the plane, we will assume that we proceed about the interior of C in a counterclockwise direction. In this case, after an edge $v_{i-1} v_i$ is encountered, the next edge encountered is $v_i v_{i+1}$, as indicated in Figure 6.8. If the color $c(v_i v_{i+1})$ of the edge $v_i v_{i+1}$ immediately follows the color $c(v_{i-1} v_i)$ of the edge $v_{i-1} v_i$ numerically (that is, if 2 follows 1, 3 follows 2, or 1 follows 3), then the vertex v_i on C is said to be of type 1. Otherwise, v_i is of type 2. See Figure 6.9 in the case where $c(v_{i-1} v_i) = 1$.

In addition to proper 3-colorings of the edges of cycles, we will be interested in labelings of the vertices of cycles where each label is either the element 1 or 2 of $\mathbb{Z}_3$. If there is such a labeling of the vertices of a cycle C such that the sum in $\mathbb{Z}_3$ of the labels of the vertices of C is 0, then the labeling is a zonal labeling of C.

Let ℓ be a labeling of the vertices of a cycle C with elements of the set $\{1, 2\}$. The labeling $\bar{\ell}$ of C defined by $\bar{\ell}(v) = 3 - \ell(v)$ for each vertex v of C is called the *complementary labeling* of ℓ. This labeling also has the property that the label of each vertex is either 1 or 2. An important characteristic of this complementary labeling is the following.

Observation 6.5 *The complementary labeling of a zonal labeling of a cycle C is also zonal. Furthermore, if a vertex v of C is of type i where $i \in \{1, 2\}$ in a proper 3-coloring c of the edges of C, then v is of type $3 - i$ in the complementary coloring $\bar{c}$ of the edges of C.*

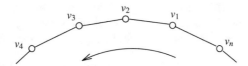

Fig. 6.8 Proceeding about a cycle in a counterclockwise direction

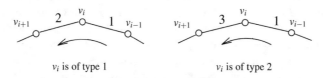

Fig. 6.9 The type of a vertex

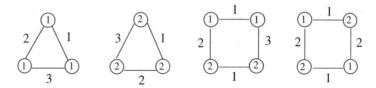

Fig. 6.10 Zonal labelings of a 3-cycle and a 4-cycle

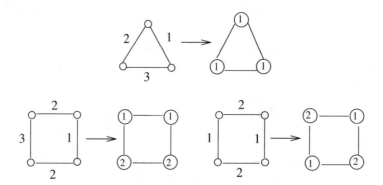

Fig. 6.11 The 3-colorings of a 3-cycle and a 4-cycle

Figure 6.10 shows possible 3-colorings of the edges of both a 3-cycle and a 4-cycle with colors from the set $\{1, 2, 3\}$, where the vertices of each cycle are labeled with their types. In each case, the resulting vertex labeling is a zonal labeling. We show that this is not a coincidence.

Theorem 6.6 *For each proper 3-coloring of the edges of an n-cycle C with colors from the set $\{1, 2, 3\}$, the resulting labeling of the vertices of C with their types is a zonal labeling of C.*

Proof We proceed by induction on the order of a cycle. Other than their complementary colorings, there is only one proper 3-coloring of the edges of a 3-cycle and essentially two proper 3-colorings of the edges of a 4-cycle. (see Figure 6.11.) In each case, the labeling of the vertices with their types is a zonal labeling.

Assume for an $(n - 1)$-cycle C', $n \geq 4$, with a proper 3-coloring of the edges of C' using the colors from the set $\{1, 2, 3\}$ that the labeling ℓ of the vertices of C' with their types is a zonal labeling. Let $C = (v_1, v_2, \ldots, v_n, v_1)$ be an n-cycle having a proper 3-coloring c of the edges of C with the colors from the set $\{1, 2, 3\}$. If only two of the three colors, say 1 and 2, are used in the coloring, then the colors 1 and 2 alternate, n is even, and the types of the vertices alternate between 1 and 2. Thus, the labeling ℓ of the vertices of C with their types is a zonal labeling. Hence, we may assume that all three colors 1, 2, and 3 are used in a proper 3-coloring of C. Necessarily, there are three consecutive edges of C having distinct colors, say

$c(v_1v_2) = 1$, $c(v_2v_3) = 2$, and $c(v_3v_4) = 3$. Then both v_2 and v_3 are of type 1. Let C' be the $(n-1)$-cycle obtained from C by identifying v_2 and v_3 and denoting the resulting vertex by v. Thus, $C' = (v_1, v, v_4, \ldots, v_n, v_1)$ is an $(n-1)$-cycle. Then $c(v_1v) = 1$ and $c(vv_4) = 3$. This gives rise to a proper 3-coloring of edges of C' where v is of type 2. All other vertices of C' have the same type as in C. By the induction hypothesis, the labeling of the vertices of C' with their types is a zonal labeling of C'. Let $s(C')$ denote the sum in $\mathbb{Z}_3$ of the vertex labels of C'. Since

$$s(C) = s(C') + \ell(v_2) + \ell(v_3) - \ell(v) = s(C') + 1 + 1 - 2 = s(C') = 0,$$

it follows that the labeling of the vertices of C with their types is also a zonal labeling. The theorem then follows by the Principle of Mathematical Induction. $\qquad\square$

The converse of Theorem 6.6 is also true.

Theorem 6.7 *For every zonal labeling ℓ of an n-cycle C_n, $n \geq 2$, there exists a proper 3-coloring of the edges of C_n such that the type of each vertex v of C_n is $\ell(v)$.*

Proof We proceed by the Strong Form of Induction on the order of a cycle. There is only one zonal labeling of C_3 (as well as its complementary zonal labeling) and two zonal labelings of C_4, as shown in Figure 6.11. Each corresponds to a proper 3-coloring of the edges of cycles (also shown in Figure 6.11.).

Assume for each zonal labeling ℓ' of a k-cycle C_k for $2 \leq k < n$, where $n \geq 5$, that there corresponds a proper 3-coloring of the edges of C_k such that for each vertex v of C_k, the type of v is $\ell'(v)$. Let $C = (v_1, v_2, \ldots, v_n, v_1)$ be an n-cycle with a zonal labeling ℓ. We show that there exists a proper 3-coloring of the edges of C such that the type of each vertex v of C is $\ell(v)$.

If each vertex of C has the same label, say 1, then $3 \mid n$ and the coloring $1, 2, 3, 1, 2, 3, \ldots$ of the edges of C in a counterclockwise direction results in each vertex of C having type 1. Hence, we may assume that both labels 1 and 2 are used in the zonal labeling ℓ of C. Therefore, there are adjacent vertices of C with distinct labels, say $\ell(v_3) = 1$ and $\ell(v_4) = 2$. Let C' be the $(n-2)$-cycle obtained from C by deleting v_3 and v_4 and joining v_2 to v_5. Then $C' = (v_1, v_2, v_5, v_6, \ldots, v_n, v_1)$ is an $(n-2)$-cycle. Let ℓ' be the labeling of C' where $\ell'(v) = \ell(v)$ for each vertex v of C'. Since ℓ is a zonal labeling of C, so is ℓ'. By the induction hypothesis, there is a proper 3-coloring of the edges of C' such that the type of each vertex v of C' is $\ell'(v)$. We may assume that $c'(v_5v_6) = 1$. We now consider four cases, depending on the labels $\ell'(v_2)$ and $\ell'(v_5)$ in C'.

> *Case 1.* $\ell'(v_2) = \ell'(v_5) = 1$. Thus, $c'(v_2v_5) = 3$ and $c'(v_1v_2) = 2$. By defining $c(v_4v_5) = 3$, $c(v_3v_4) = 1$, and $c(v_2v_3) = 3$, we obtain a proper 3-coloring of the edges of C such that the type of each vertex v of C is $\ell(v)$.
> *Case 2.* $\ell'(v_2) = \ell'(v_5) = 2$. Thus, $c'(v_2v_5) = 2$ and $c'(v_1v_2) = 3$. By defining $c(v_4v_5) = 2$, $c(v_3v_4) = 3$, and $c(v_2v_3) = 2$, we obtain a proper 3-coloring of the edges of C such that the type of each vertex v of C is $\ell(v)$.

Case 3. $\ell'(v_2) = 1$ and $\ell'(v_5) = 2$. Thus, $c'(v_2v_5) = 2$ and $c'(v_1v_2) = 1$. By defining $c(v_4v_5) = 3$, $c(v_3v_4) = 1$, and $c(v_2v_3) = 3$, we obtain a proper 3-coloring of
the edges of C such that the type of each vertex v of C is $\ell(v)$.
Case 4. $\ell'(v_2) = 2$ and $\ell'(v_5) = 1$. Thus, $c'(v_2v_5) = 3$ and $c'(v_1v_2) = 1$. By defining $c(v_4v_5) = 3$, $c(v_3v_4) = 1$, and $c(v_2v_3) = 3$, we obtain a proper 3-coloring of
the edges of C such that the type of each vertex v of C is $\ell(v)$.   □

Corollary 6.1 *For every zonal labeling ℓ of an n-cycle C_n, $n \geq 2$, and for a color
a from the set $\{1, 2, 3\}$ assigned to an arbitrary edge e of C_n, there exists a unique
proper 3-coloring c of the edges of C_n such that $c(e) = a$ and the type of each vertex
v of C_n is $\ell(v)$.*

This brings up a fundamental question.

Problem 6.1 Which connected plane graphs are zonal?

6.3   Zonal Labelings of Cubic Maps

We now turn our attention back to cubic maps. Let M be a cubic map, that is, M is a
connected bridgeless cubic planar graph (or multigraph) embedded in the plane. We
know by the Four Color Theorem that there is a proper coloring of the edges of M
with three colors, say 1, 2, 3. Let such a coloring be given. We now define the type
of a vertex in M in which a proper 3-coloring of its edges has been given using the
colors 1, 2, 3. Let v be a vertex of M. There are three edges incident with v and these
three edges are colored 1, 2, 3 in one of two ways as we proceed clockwise about v.
Depending on the order in which the colors 1, 2, 3 are encountered, v is either type 1
or type 2. The vertex v is type 1 if the colors are encountered in the order 1-2-3 and
v is type 2 if the colors are encountered in the order 1-3-2. See Figure 6.12.

Let there be given a proper 3-coloring of the edges of a cubic map M and label each
vertex of M either 1 or 2, according to its type. For example, the 3-cube $Q_3 = C_4 \,\square\, K_2$
is a cubic map. A proper 3-coloring of the edges of Q_3 with the colors 1, 2, 3 is shown
in Figure 6.13a. The vertices of Q_3 are also labeled with their types as obtained from
this 3-coloring. This labeling is shown in Figure 6.13b. It turns out that this labeling
of the vertices of Q_3 is a zonal labeling as the label of every region is 0.

Fig. 6.12 The vertex types
in a proper 3-coloring of the
edges of a cubic map

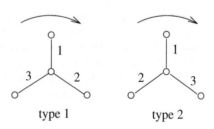

type 1   type 2

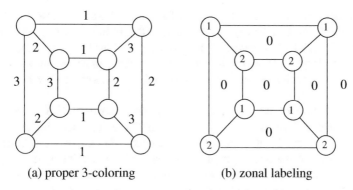

(a) proper 3-coloring   (b) zonal labeling

Fig. 6.13   A proper 3-coloring of the edges and a zonal labeling of Q_3

What we have seen is that a certain proper 3-coloring of the edges of the cubic map Q_3 has given rise to a labeling of the vertices of Q_3 with their types, and this proved to be a zonal labeling of Q_3. This too is no coincidence. The following result is the analogue of Theorem 6.6 for cubic maps.

Theorem 6.8   *For every proper 3-coloring of the edges of a cubic map M with the colors $1, 2, 3$, the labeling of the vertices of M with their types is a zonal labeling of M.*

Proof   Let R be a zone of M and let C be the boundary cycle of R. Now, let there be given a proper 3-coloring of the edges of M with the colors $1, 2, 3$. First, observe that the type of a vertex v on C according to the colors of its three incident edges is the same as the type of v for the coloring of the edges of C. By Theorem 6.6, the labeling of the vertices of C with their types is a zonal labeling of C. Since this is the case for every boundary cycle, this labeling of the vertices of M is a zonal labeling of M.   □

There is also a result for cubic maps that is the analogue of Theorem 6.7. First, we present a useful observation.

Observation 6.9   *Let G be a connected cubic plane graph with a zonal labeling and let C be the boundary cycle of a region of G. Then the zonal labeling restricted to the vertices of C is a zonal labeling of C.*

Theorem 6.10   *Let M be a cubic map with a zonal labeling ℓ. Then there exists a proper 3-coloring of the edges of M with the colors $1, 2, 3$ such that the type of each vertex x of M is $\ell(x)$. Furthermore, if some edge of M is assigned one of the colors $1, 2, 3$, then a proper 3-coloring of the edges of M is uniquely determined.*

Proof   Let M be a cubic map with a zonal labeling ℓ and let e be an edge of M, where e is assigned a color $c(e)$ from the set $\{1, 2, 3\}$. Let u be a vertex of M that is incident with e. According to the label $\ell(u)$ of u, we assign the colors $1, 2, 3$ to the three edges incident with u such that the type of u is $\ell(u)$. Next, we show that the color $c(e)$ of e and the zonal labeling ℓ of M uniquely determines a proper 3-coloring c of the edges of M such that the type of each vertex x of M is $\ell(x)$. Let f be an arbitrary edge of M and suppose that f is incident with a vertex v of M. We may assume that f is not incident with u. Since M is 2-connected, there is a cycle C of M containing both u and v.

First, suppose that C is the boundary cycle of some zone of M. By Observation 6.9, the zonal labeling ℓ restricted to the vertices of C is a zonal labeling of C. It then follows by Corollary 6.1 that the colors of all edges of C are uniquely determined. Thus, the colors of the two edges of C incident with v are uniquely determined, as is the color of the edge f. Hence, we may assume that C is not the boundary cycle of any zone of M. Therefore, C encloses more than one zone of M.

Let $C^{(1)}$ be the boundary cycle of a zone $R^{(1)}$ of M lying within C such that u lies on $C^{(1)}$. The restriction of the zonal labeling ℓ to the vertices of $C^{(1)}$ is a zonal labeling of $C^{(1)}$. By Observation 6.9 and Corollary 6.1, the colors of the edges of $C^{(1)}$ are uniquely determined. There exists a sequence $C^{(1)}, C^{(2)}, \ldots, C^{(k)}$ of cycles, $k \geq 2$, where $C^{(i)}$ is the boundary cycle of a zone $R^{(i)}$ lying within C such that $C^{(i)}$ and $C^{(i+1)}$ have an edge in common for $1 \leq i \leq k-1$ and v is a vertex on $C^{(k)}$. Since the colors of the edges of $C^{(1)}$ are uniquely determined and an edge of $C^{(1)}$ belongs to $C^{(2)}$, the colors of the edges of $C^{(2)}$ are uniquely determined. Continuing in this manner, we see that the colors of the edges of $C^{(k)}$ are uniquely determined as well, as is the color of the edge f. □

As a consequence of Theorem 6.10, it follows that for every cubic map M possessing a zonal labeling, there exists a proper 3-coloring of the edges of M. This in turn implies the Four Color Theorem. Therefore, to give an alternative proof of the Four Color Theorem, it suffices to prove that every cubic map M has a zonal labeling. Of course, since the Four Color Theorem is known to be true, we know that there is a proper 3-coloring of the edges of every cubic map. Therefore, the following corollary is a consequence of Theorem 6.8.

Corollary 6.2 *Every cubic map has a zonal labeling.*

Theorems 6.3 (Tait's theorem), 6.10, and Corollary 6.2 give rise to the following theorem.

Theorem 6.11 *There exists a proper 4-coloring of the zones of a cubic map M if and only if M has a zonal labeling.*

By Corollary 6.2, one direction of Conjecture 6.1 is true, namely, if a connected cubic plane graph G is bridgeless (that is, if G is a cubic map), then G is zonal. We now show that this conjecture is true in its entirety. It is useful to recall a property possessed by graphs of odd order having a sufficiently large size. A graph G of odd order n and size m is called *overfull* if $m > \Delta(G)(n-1)/2$. The following result is well known (see [9, p. 258], for example).

Theorem 6.12 *If G is an overfull graph, then G does not possess a proper $\Delta(G)$-coloring of its edges.*

The following argument is similar to the one used in the proof of Theorem 6.10.

Theorem 6.13 *If G is a connected cubic plane graph with bridges, then G is not zonal.*

Proof Assume, to the contrary, that there is a connected cubic plane graph G with bridges possessing a zonal labeling ℓ. Let $e = uv$ be a bridge of G such that v belongs to an end-block B of G. Necessarily, the order of B is at least 5 and v is incident with two edges of B, say vx and vy. According to the label $\ell(v)$ of v, we may assign the colors 1, 2, 3 to the three edges incident with v such that the type of v is $\ell(v)$. Next, we show that the color $c(e)$ of e and the zonal labeling ℓ of G uniquely determines a proper 3-coloring c of the edges of B such that the type of each vertex v' of B is $\ell(v')$. Let f be an arbitrary edge of B where f is incident with a vertex w of B. Since the color of each edge incident with v is uniquely determined by $c(e)$ and $\ell(v)$, we may assume that f is not incident with v. Since B is 2-connected, there is a cycle C of B containing both v and w. Then xv and vy are two adjacent edges of C.

First, suppose that C is the boundary cycle of some zone of B. By Observation 6.9, the zonal labeling ℓ restricted to the vertices of C is a zonal labeling of C. Since the color of the edge vx of C is given, it then follows by Corollary 6.1 that the colors of all edges of C are uniquely determined. Thus, the colors of the two edges of C incident with w are uniquely determined, as is the color of the edge f. Hence, we may assume that C is not the boundary cycle of any zone of B. Therefore, C encloses more than one zone of B.

Let $C^{(1)}$ be the boundary cycle of a zone $R^{(1)}$ of B lying within C such that v lies on $C^{(1)}$. The zonal labeling ℓ restricted to the vertices of $C^{(1)}$ is a zonal labeling of $C^{(1)}$. By Observation 6.9 and Corollary 6.1, the colors of the edges of $C^{(1)}$ are uniquely determined. There exists a sequence $C^{(1)}, C^{(2)}, \ldots, C^{(k)}$ of cycles, $k \geq 2$, where $C^{(i)}$ is the boundary cycle of a zone $R^{(i)}$ lying within C such that $C^{(i)}$ and $C^{(i+1)}$ have an edge in common for $1 \leq i \leq k - 1$ and w is a vertex on $C^{(k)}$. Since the colors of the edges of $C^{(1)}$ are uniquely determined and an edge of $C^{(1)}$ belongs to $C^{(2)}$, the colors of the edges of $C^{(2)}$ are uniquely determined. Continuing in this manner, we see that the colors of the edges of $C^{(k)}$ are uniquely determined as well, as is the color of the edge f. Thus, there is a proper 3-coloring of the edges of B.

Let $p \geq 5$ be the order of B. Since B has $p - 1$ vertices of degree 3 and one vertex of degree 2, it follows that p is odd and the size of B is $\frac{3p-1}{2} > \frac{3(p-1)}{2} = \frac{\Delta(B)(p-1)}{2}$ and so B is a overfull graph. By Theorem 6.12, there is no proper 3-coloring of the edges of B. This is a contradiction.   □

The following corollary is a consequence of Corollary 6.2 and Theorem 6.13, which establishes the truth of Conjecture 6.1.

Theorem 6.14 *A connected cubic plane graph G is zonal if and only if G is bridgeless.*

6.4   Inner Zonal Labelings of Bicubic Maps

The argument given that establishes the truth of Theorem 6.11 makes use of the fact that we knew that the Four Color Theorem is true. It would be considerably more interesting if it could be shown that every cubic map is zonal without using the Four Color Theorem (and without using computers). We state this fundamental question next.

Problem 6.2   Can every cubic map be proved to be zonal without using the Four Color Theorem or computers?

Of course, we were able to establish the zonality of cycles without using the Four Color Theorem. Every cycle is a subgraph and the boundary of a zone of some cubic map. This brings up the question of the existence of certain types of labelings of graphs that are subgraphs of some cubic map.

By a *bicubic map B* is meant a 2-connected planar graph embedded in the plane all of whose vertices have degree 2 or 3, where the boundary of the exterior region of B is a cycle C and every vertex lying interior to C has degree 3. Therefore, if C is a cycle in a cubic map M and H is the submap of M consisting of C and all vertices and edges of M lying interior to C, then H is a bicubic map. If no vertex lies interior to C, then either $H = C$ or H is C with chords; while if every vertex of H has degree 3, then $H = M$. This suggests a labeling closely related to a zonal labeling.

Let B be a bicubic map. A labeling of the vertices of B with the elements 1 and 2 of $\mathbb{Z}_3$ is an *inner zonal labeling* if the label of each interior zone of B is 0. If B contains an inner zonal labeling, then B is *inner zonal*. Therefore, the labeling of $K_4 - e$ given in Figure 6.4 is an inner zonal labeling and so $K_4 - e$, while not zonal, is inner zonal. Of course, if a bicubic map B has an inner zonal labeling such that the label of the exterior zone is 0 as well, then the labeling is zonal and B itself is zonal. Therefore, if B is a cycle or a cubic map, then B is inner zonal. The question is which bicubic maps are inner zonal and which of these can be shown to have this property—without using the Four Color Theorem or computers.

There is an observation for inner zonal labelings of bicubic maps that is the analogue of Observation 6.1.

Observation 6.15   *If ℓ is an inner zonal labeling of a bicubic map, then so too is its complementary labeling $\overline{\ell}$.*

Lemma 6.1   *Let C be an n-cycle, $n \geq 4$, and let $u_1 v_1$ and $u_2 v_2$ be two nonadjacent edges of C. There exists a zonal labeling ℓ of C such that $\ell(u_i) = \ell(v_i)$ for $i = 1, 2$.*

Proof If $n = 2k \geq 4$, then we may label k vertices of C (including u_1 and v_1) the color 1 and the remaining k vertices of C (including u_2 and v_2) the color 2. If $n = 2k + 1 \geq 5$, then we may label $k + 2$ vertices of C (including $u_1, v_1, u_2,$ and v_2) the color 1 and the remaining $k - 1$ vertices of C the color 2. In each case, we have a zonal labeling of C. $\square$

Actually, every bicubic map is inner zonal. To see this, let G be a bicubic map and let G' be another copy of G embedded in the plane in the same way that G is. For each vertex v of G, let v' be the vertex of G' corresponding to v. Then a plane graph H can be constructed from G and G' by adding the edge uu' for each vertex u of G having degree 2. Thus, H is a cubic map and by Theorem 6.8, H has a zonal labeling ℓ. This labeling ℓ, restricted to G, is an inner zonal labeling of G. However, the reason we know that H has a zonal labeling is because there is a proper 3-coloring of the edges of H and the reason we know this is because the Four Color Theorem is true. And, the reason we know that the Four Color Theorem is true is because there is a computer-aided proof of this theorem. This brings up the question as to whether there exists an independent proof of the inner zonality of every bicubic map, namely, one that does not require prior knowledge of the Four Color Theorem. We now give such a proof for one class of bicubic maps.

Theorem 6.16   *Let G be a bicubic map where the boundary of the exterior region is a Hamiltonian cycle C with k pairwise nonadjacent chords for some positive integer k. There exists an inner zonal labeling ℓ of G where $\ell(x) = \ell(y)$ for each chord xy of C.*

Proof We proceed by induction on k. First, suppose that G is a bicubic map where the boundary of the exterior region is a Hamiltonian cycle of G with one chord xy. Let R_1 and R_2 be the two interior zones of G. Label x and y with the color 1. If the boundary C_1 of R_1 is a triangle, then we label the remaining vertex on C_1 the color 1. If C_1 is an m-cycle for some integer $m \geq 4$, then, by Lemma 6.1 and Observation 6.15, there exists a zonal labeling ℓ_1 of C_1 such that $\ell_1(x) = \ell_1(y) = 1$. Similarly, there is a zonal labeling ℓ_2 of the boundary of R_2 such that $\ell_2(x) = \ell_2(y) = 1$. We now define an inner zonal labeling ℓ of G by $\ell(w) = \ell_1(w)$ if $w \in V(C_1)$ and $\ell(w) = \ell_2(w)$ if $w \in V(G) - V(C_1)$. Then $\ell(x) = \ell(y)$. Thus, the basis step of the induction holds.

Assume that the statement is true for all bicubic maps in which the boundary of the exterior region is a Hamiltonian cycle with k pairwise nonadjacent chords for some positive integer k. Let G be a bicubic map in which the boundary of the exterior region is a Hamiltonian cycle C with $k + 1$ pairwise nonadjacent chords. Let R_1 be an interior zone of G such that the boundary C_1 of R_1 contains only one chord uv of C. Deleting $V(C_1) - \{u, v\}$ from G produces a bicubic map G' in which the boundary of the exterior region is a Hamiltonian cycle C' with k pairwise nonadjacent chords. By the induction hypothesis, there exists an inner zonal labeling ℓ' of G' such that $\ell'(s) = \ell'(t)$ for each chord st of C'. Next, we show that the bicubic map G has an inner zonal labeling ℓ such that $\ell(p) = \ell(q)$ for every chord pq of C. We consider two cases, according to whether $\ell'(u) = \ell'(v)$ or $\ell'(u) \neq \ell'(v)$.

Case 1. $\ell'(u) = \ell'(v)$. By Observation 6.15, we may assume that $\ell'(u) = \ell'(v) = 1$. If C_1 is a triangle, then we label the remaining vertex of C_1 the color 1. Thus, we may assume that C_1 is an m-cycle, where $m \geq 4$. By Lemma 6.1 and Observation 6.15, there exists a zonal labeling ℓ_1 of C_1 such that $\ell_1(u) = \ell_1(v) = 1$. Then an inner zonal labeling ℓ of G can be constructed from the labelings ℓ' and ℓ_1 by defining $\ell(w) = \ell_1(w)$ if $w \in V(C_1)$ and $\ell(w) = \ell'(w)$ otherwise.

Case 2. $\ell'(u) \neq \ell'(v)$. Let R_2 be the interior zone of G that is adjacent to R_1. Thus, R_2 contains uv on its boundary as well as another chord xy of G. It then follows by the defining property of ℓ' that $\ell'(x) = \ell'(y)$. Let C_2 be the boundary cycle of R_2. Then uv and xy are two nonadjacent edges of C_2. By Lemma 6.1 and Observation 6.15, there exists a zonal labeling ℓ_2 of C_2 such that $\ell_2(u) = \ell_2(v)$ and $\ell_2(x) = \ell_2(y) = \ell'(x) = \ell'(y)$. We now define another inner zonal labeling ℓ^* of G' by $\ell^*(w) = \ell_2(w)$ if $w \in V(C_2)$ and $\ell^*(w) = \ell'(w)$ if $w \in V(G') - V(C_2)$. Then ℓ^* is an inner zonal labeling of G' where $\ell^*(s) = \ell^*(t)$ for each chord st of C' and $\ell^*(u) = \ell^*(v)$. We now proceed as in Case 1 to construct an inner zonal labeling ℓ of G.

In each case, the inner zonal labeling ℓ of G has the property that $\ell'(s) = \ell'(t)$ for each chord st of C. Therefore, the theorem follows by the Principle of Mathematical Induction.   □

By Theorem 6.16, every bicubic map G in which the boundary of the exterior region is a Hamiltonian cycle C is inner zonal. If $G = C$, then G is zonal. Thus, we have the following corollary, the truth of which is independent of the Four Color Theorem.

Corollary 6.3   *Every plane graph G with $\Delta(G) \leq 3$ where the boundary cycle of the exterior zone is a Hamiltonian cycle of G is inner zonal.*

References

1. M. Aigner, E. Triesch, Irregular assignments of trees and forests. SIAM J. Discret. Math. **3**, 439–449 (1990)
2. L.W. Beineke, Characterizations of derived graphs. J. Comb. Theory **9**, 129–135 (1970)
3. C. Berge, *Théorie des Graphes et Ses Applications* (Dunod, Paris, 1958)
4. N.L. Biggs, E.K. Lloyd, R.J. Wilson, *Graph Theory 1736–1936* (Clarendon Press, Oxford, 1976)
5. A. Cayley, A theorem on trees. Q. J. Math. **23**, 376–378 (1889)
6. G. Chartrand, M.S. Jacobson, J. Lehel, O.R. Oellermann, S. Ruiz, F. Saba, Irregular networks. Congr. Numer. **64**, 197–210 (1988)
7. G. Chartrand, F. Okamoto, P. Zhang, The sigma chromatic number of a graph. Graphs Comb. **26**, 755–773 (2010)
8. G. Chartrand, L. Lesniak, P. Zhang, *Graphs & Digraphs*, 6th edn. (Chapman & Hall/CRC, Boca Raton, FL, 2015)
9. G. Chartrand, P. Zhang, *Chromatic Graph Theory* (Chapman & Hall/CRC Press, Boca Raton, 2009)
10. T. Deretsky, S.M. Lee, J. Mitchem, On vertex prime labelings of graphs, in *Graph Theory, Combinatorics, and Applications* (Wiley-Interscience, New York, 1991), pp. 359–369
11. P. Erdős, R.J. Wilson, On the chromatic index of almost all graphs. J. Comb. Theory Ser. B **23**, 255–257 (1977)
12. W. Fang, A computational approach to the graceful tree conjecture (2010). arXiv: 1003.3045v2
13. R.J. Faudree, M.S. Jacobson, J. Lehel, R.H. Schelp, Irregular networks, regular graphs and integer matrices with distinct row and column sums. Discret. Math. **76**, 223–240 (1989)
14. M. Fiedler (ed.), Theory of Graphs and its Applications, in *Proceedings of the Symposium Held in Smolenice in June 1963*. Prague, Pub. House of the Czechoslovak Academy of Sciences (New York, Academic Press, 1964)
15. R. Frucht, Herstellung von Graphen mit vorgegebener abstrakter Gruppe. Compositio Math. **6**, 239–250 (1938)
16. R. Frucht, Graceful numbering of wheels and related graphs. Ann. N. Y. Acad. Sci. **319**, 219–229 (1979)
17. H.L. Fu, K.C. Huang, On prime labellings. Graph Theory Appl. Discret. Math. **127**, 181–186 (1994)
18. J.A. Gallian, A dynamic survey of graph labeling. Electron. J. Comb. (2017) (#DS6)
19. S.W. Golomb, How to number a graph, *Graph Theory and Computing* (Academic Press, New York, 1972), pp. 23–37
20. R.L. Graham, N.J.A. Sloane, On additive bases and harmonious graphs. SIAM J. Alg. Disc. Math. **1**, 382–404 (1980)
21. A. Gyárfás, The irregularity strength of $K_{m,m}$ is 4 for odd m. Discret. Math. **71**, 273–274 (1988)

© The Author(s), under exclusive license to Springer Nature Switzerland AG 2019
G. Chartrand et al., *How to Label a Graph*, SpringerBriefs
in Mathematics, https://doi.org/10.1007/978-3-030-16863-6

22. P. Haxell, O. Pikhurko, A. Taraz, Primality of trees. J. Comb. **2**, 481–500 (2011)
23. P. Hrnčiar, A. Haviar, All trees of diameter five are graceful. Discret. Math. **233**, 133–150 (2001)
24. C. Huang, A. Kotzig, A. Rosa, Further results on tree labellings. Util. Math. **21**, 31–48 (1982)
25. M. Kalkowski, M. Karoński, F. Pfender, Vertex-coloring edge-weightings: towards the 1-2-3-conjecture. J. Comb. Theory Ser. B **100**, 347–349 (2010)
26. M. Karoński, T. Łuczak, A. Thomason, Edge weights and vertex colours. J. Comb. Theory Ser. B **91**, 151–157 (2004)
27. M. Kneser, Aufgabe 300. Jahresber. Deutsch., Math. Verein. **58**, 27 (1955)
28. D. König, *Theorie der endlichen und unendliehen Graphen* (Akademische Verlagsgesellschaft, Leipzig, 1936)
29. A. Kotzig, A. Rosa, Magic valuations of finite graphs. Canad. Math. Bull. **13**, 451–461 (1970)
30. S.M. Lee, I. Wui, J. Yeh, On the amalgamation of prime graphs. Bull. Malays. Math. Soc. **11**, 59–67 (1988)
31. L. Lovász, Kneser's conjecture, chromatic number, and homotopy. J. Comb. Theory Ser. A **25**, 319–324 (1978)
32. T.A. McKee, F.R. McMorris, *Topics in Intersection Graph Theory*, SIAM Monographs on Discrete Mathematics and Applications (Philadelphia, PA, 1999)
33. J.W. Moon, Counting labeled trees, in *From lectures delivered to the Twelfth Biennial Seminar of the Canadian Mathematical Congress (Vancouver, 1969)* (1970). (Canadian Mathematical Monographs, Canad. Math. Congress, Montreal, Que)
34. O. Ore, *Theory of Graphs* (Providence, RI, 1962). (Am. Math. Soc. Colloq. Pub)
35. J. Petersen, Die Theorie der regulären Graphen. Acta Math. **15**, 193–220 (1891)
36. H. Prüfer, Neuer Beweis eines Satzes über Permutationen. Arch. Math. Phys. **27**, 142–144 (1918)
37. S.N. Rao, Prime labeling, in *R. C. Bose Centenary Symposium on Discrete Mathematics. and Applications, Kolkata* (2002)
38. G. Ringel, Problem 25, in *Theory of Graphs and its Applications* (Nakl. ČSAV, Prague, 1964), p. 162
39. G. Ringel, J.W.T. Youngs, Solution of the Heawood map-coloring problem. Proc. Nat. Acad. Sci. USA **60**, 438–445 (1968)
40. A. Rosa, On certain valuations of the vertices of a graph, *Theory of Graphs* (Gordon and Breach, New York, 1967), pp. 349–355
41. J. Sedláček, Problem #27, Theory of Graphs and its Applicationsin, *Proceedings of the Symposium Held in Smolenice in June 1963*. Prague, Pub. House of the Czechoslovak Academy of Sciences; (New York, Academic Press, 1964)
42. J. Sedláček, On magic graphs. Math. Slov. **26**, 329–335 (1976)
43. M.A. Seoud, A.T. Diab, E.A. Elsakhawi, On strongly-C harmonious, relatively prime, odd graceful and cordial graphs. Proc. Math. Phys. Soc. Egypt **73**, 33–55 (1998)
44. S. Stahl, *n*-Tuple colorings and associated graphs. J. Comb. Theory Ser. B **20**, 185–203 (1976)
45. M.A. Seoud, M.Z. Youssef, On prime labelings of graphs. Congr. Numer. **141**, 203–215 (1999)
46. N.J.A. Sloane, *The On-Line Encyclopedia of Integer Sequences* (OEIS). Sequence Number A213273
47. B.M. Stewart, Magic graphs. Canad. J. Math. **18**, 1031–1059 (1966)
48. B.M. Stewart, Supermagic complete graphs. Canad. J. Math. **19**, 427–438 (1967)
49. P.G. Tait, Remarks on the colouring of maps. Proc. Royal Soc. Edinburgh **10**(729), 501–503 (1880)
50. R. Tout, A.N. Dabboucy, K. Howalla, Prime labeling of graphs. Nat. Acad. Sci. Lett. **5**, 365–368 (1982)
51. V.G. Vizing, On an estimate of the chromatic class of a *p*-graph. (Russian) *Diskret. Analiz.* **3**, 25–30 (1964)
52. R. Wilson, *Four Colors Suffice: How the Map Problem Was Solved* (Princeton University Press, Princeton, NJ, 2002)
53. P. Zhang, *Color-Induced Graph Colorings* (Springer, New York, 2015)
54. P. Zhang, *A Kaleidoscopic View of Graph Colorings* (Springer, New York, 2016)

Index

© The Author(s), under exclusive license to Springer Nature Switzerland AG 2019
G. Chartrand et al., *How to Label a Graph*, SpringerBriefs
in Mathematics, https://doi.org/10.1007/978-3-030-16863-6